Lecture Notes in Mathematics

Edited by A. Dold and B. Eckmann

837

Josef Meixner
Friedrich W. Schäfke
Gerhard Wolf

Mathieu Functions and Spheroidal Functions and Their Mathematical Foundations

Further Studies

Springer-Verlag
Berlin Heidelberg New York 1980

Authors

Josef Meixner
Am Blockhaus 31
5100 Aachen
Federal Republic of Germany

Friedrich W. Schäfke
Fakultät für Mathematik
Universität Konstanz
Postfach 5560
7750 Konstanz
Federal Republic of Germany

Gerhard Wolf
FB 6 Mathematik
Universität-Gesamthochschule
Universitätsstraße 3
Postfach 6843
4300 Essen 1
Federal Republic of Germany

AMS Subject Classifications (1980): 33A40, 33A45, 33A55, 34A20, 34B25, 34B30, 34D05, 34E05, 35J05, 47A70

ISBN 3-540-10282-5 Springer-Verlag Berlin Heidelberg New York
ISBN 0-387-10282-5 Springer-Verlag New York Heidelberg Berlin

Printed in Germany

Printing and binding: Beltz Offsetdruck, Hemsbach/Bergstr.
2141/3140-543210

TABLE OF CONTENTS

Introduction and Preface.

1. Foundations. 1

1.1. Eigenvalue problems with two parameters. 1

1.1.0. Introduction. 1

1.1.1. First presuppositions. Preliminary remarks. 3

1.1.2. Estimates for the resolvent. 7

1.1.3. The eigenvalues to $\mu \neq 0$. 11

1.1.4. Further presuppositions and conclusions 15

1.1.5. The residues of the resolvent. Principal solutions. 16

1.1.6. Equiconvergence. 22

1.1.7. Holomorphy properties. Estimates. 23

1.1.8. Additional estimates. 26

1.1.9. On the application to boundary value problems for ordinary differential equations and differential systems. 27

1.1.10. Application to Hill's differential equation in the real domain. 33

1.1.11. Application to Hill's differential equation in the complex domain. 36

1.1.12. Application to the spheroidal differential equation in the real domain. 39

1.1.13. Application to the spheroidal differential equation in the complex domain. 42

1.2. Simply separated operators. 44

1.2.0. Introduction. 44

1.2.1. The algebraic problem. 46

1.2.2. Adjoint mappings. 50

1.2.3. The analytical problem. Expansion theorem. 53

1.2.4. The symmetric case. 58

1.2.5. Applications. 60

2. Mathieu Functions. 63

2.1. Integral relations. 63

2.1.1. Integral relations of the first kind. 63

2.1.2. Integral relations of the second kind (with variable boundaries). 71

2.2. Addition theorems. 73

2.2.1. Lemmas concerning the transformation equation. 74

2.2.2. Integral relations. 75

2.2.3. The addition theorems. 77

2.2.4. Consequences and special cases. 80

2.3. On the computation of the characteristic exponent. 83

2.4. On the eigenvalues for complex h^2. 85

2.5. Improved estimates of the radii of convergence. 90

2.6. Asymptotic estimates for large h^2. 96

2.7. On the power series of the eigenvalues. 99

3. Spheroidal Functions. 102

3.1. Integrals with products of spheroidal functions. 102

3.1.1. Integral relations of the first kind. 102

3.1.2. Integral relations of the second kind. 104

3.2. On the eigenvalues for complex γ^2. 106

3.3. The spheroidal functions for $\mu^2 = 1$, $\lambda = 0$. 110

3.4. Applications and numerical tables. 113

Appendix. Corrections of errors in MS. 117

Bibliography. 120

Introduction and Preface

More than 20 years have passed since the publication of the first books on Mathieu functions and spheroidal functions by MacLachlan (1947), Flammer (1952), Stratton, Morse, Chu, Little, Corbató (1956), Campbell (1955) and by Meixner and Schäfke (1954). In this period, the field has seen essential progress in various directions.

On the one hand the advent of computers has greatly advanced the numerical mastery of the functions and of the corresponding eigenvalues, which is so important for practical applications. Thus extensive and voluminous relevant tables now exist. A similar, practical goal is pursued in numerous papers in which various kinds of asymptotic formulas are derived.

On the other hand there has been much progress in the mathematical theory, going beyond even the comprehensive book by Meixner and Schäfke. The present account is devoted to the most important aspects of this progress. It follows in formulation and notations the book by Meixner and Schäfke, which is quoted in the following by MS.

It is well known that Mathieu functions and spheroidal functions are the simplest classes of special functions of mathematical physics which arise from the separation of the (2-or) 3-dimensional time independent wave equation and which are not essentially hypergeometric functions. The separation of the wave equation $\Delta u + k^2 u = 0$ in the coordinates of the elliptic cylinder or in spheroidal coordinates, respectively, yields in essence two identical ordinary differential equations. In their rational form they possess two finite regular singularities and the non-regular singularity ∞. These differential equations are coupled by a separation parameter λ and also contain the parameter h^2, respectively γ^2, which combines the parameter k and the parameter of the coordinate system. The theory of Mathieu and spheroidal functions as well as its main problems stem from this origin. Thus non-trivial 2-parametric eigenvalue problems play a fundamental part. They lead, as first shown by Meixner and Schäfke in MS, to expansion theorems in terms of Mathieu functions and spheroidal functions and therewith to the expansions of solutions of the wave equation in terms of products of the mentioned functions, in particular to addition theorems.

Hitherto these theorems were formulated and proved throughout only for the case of "normal" values of h^2 or γ^2, and accordingly not for all values of k^2: "exceptional values" of h^2 or γ^2, for which eigenvalues coalesce, were excluded. The considerations of 1. are, among other points, devoted to this gap. In 1.1. a very comprehensive theory of two-parametric eigenvalue problems is presented. In particular it furnishes, both in the real and in the complex domain, expansion theorems and asymptotic formulas for the exceptional points as well. For a description of this theory we must refer to the introduction 1.1.0. .

The significance of the exceptional points for the representation of the solutions of the wave equation with an exceptional value k^2 will be shown very generally in 1.2. for a class of "simply separable operators". Here, in a natural way, tensor products play an important part. We have, however, largely foregone an explicit notation of the definite specializations for Mathieu functions and spheroidal functions; this holds in particular for the addition theorems.

The fact that the two Mathieu differential equations, which arise in the separation of the wave equation, are coupled leads to integral relations for the Mathieu functions. This is well known; but during the last few years some interesting aspects and remarkable new integral relations have emerged. Section 2.1. is devoted to them. It was already shown in the book by Meixner and Schäfke that the "addition theorem of the Mathieu functions" can play a central part in the theory and that it contains as special cases most of the important series expansions. It is true that its former formulation has two disadvantages: it is only an "exterior" theorem, and the coefficients are expressed not in elliptical but in polar coordinates. Here the investigations of 2.2. come in and lead to a completely satisfactory general theorem. It turns out, that the "exterior" and the "interior" theorems have, apart from a simple rearrangement of the series, the same form if the coefficients are also expressed in elliptical coordinates. The investigations are performed for the whole complex domain of validity.

One recognizes from the nature of the differential equations, as described above, that the circuital behavior of the solutions around ∞ cannot just be read off from the indices of the simple singularities. Thus the determination of the corresponding characteristic exponents is a fundamental problem. In 2.3. it is described how the three term recursions of the Fourier series coefficients of the Mathieu functions can be used to obtain a simple and direct computational method.

Section 2.4. is again, and more thorougly, devoted to the eigenvalues for complex values of h^2, and, of course, to the branch points which occur for those complex h^2, whose h^2-projections are the exceptional points. For the first time an extensive table of computed branch points is imparted. It leads to well founded conjectures on the asymptotic distribution of such points and consequently on the radii of convergence of the function elements about $h^2 = 0$. It is remarkable that the radii of convergence seem to grow with the eigenvalue number according to a quadratic law while the lower estimates given by MS using perturbation theoretical methods are linear. A proof and a theory which leads to a deeper understanding of this fact are still missing. However, another open problem can now be answered. In a remarkably simple way one can prove that, for instance, the function elements $a_{2n}(h^2)$ about $h^2 = 0$ all belong to o n e analytical function in the large, that is they follow from each other by analytical continuation.

Incidentally we remark that the good numerical knowledge of the radii of convergence is useful for a better utilization of the power series of the eigenvalues for the purpose of their numerical computation.

In 2.5. we present, as further development of an idea due to A.Schönhage, a method for the derivation of improved lower estimates of the radii of convergence of the eigenvalues. For the $\lambda_{\nu+2m}(h^2)$ with $\nu \in \mathbb{C} \setminus \mathbb{Z}$ as well as for the four classes with $\nu = 0$ or $\nu = 1$ one can give bounds which increase proportionally to $|m| \cdot \log |m|$.

In 2.6. we note error estimates for the asymptotics of eigenvalues and eigensolutions $a_n(h^2)$, $ce_n(z,h^2)$ $(n \in \mathbb{N}_0)$ which - with much work, tenacity and skill - have been obtained by M.Kurz. Estimates of this kind have not been known before. This is apparently due, mainly, to the remarkable difficulties of execution encountered.

In 2.7. the link between the power series expansions of the $\lambda_{\nu+2n}(h^2)$ for non-integer ν with those of the $a_n(h^2)$, $b_n(h^2)$ is more closely studied. The method also applies to the corresponding functions.

Two types of integrals with products of spheroidal functions are examined in 3.1.. In 3.1.1. integrals with products of spheroidal functions are directly obtained from the spheroidal differential equation. In particular, certain orthogonality relations result. In 3.1.2. the fact is used that Δ commutes with the derivatives with respect to the cartesian variables. It was already applied to Mathieu functions in MS. Also in the present case no recursions as with the simple special functions result, but just a special kind of integral relations.

In 3.2. the eigenvalues of the spheroidal differential equation for complex γ^2 are studied more thoroughly. In particular, the extensive tables of branch points are important and, also here, lead to conjectures on their distribution. Thus the radii of convergence about 0 seem to increase with the square of the eigenvalue number. As in 2.4. one can in many cases demonstrate the irreducibility, which means that certain function elements about 0 belong to o n e function in the large. Again the numerical knowledge of the radii of convergence can be used for an improved application of the power series of the eigenvalues for their computation.

In 3.3. we study the interesting case $\mu^2 = 1$, $\lambda = 0$. In 3.4. we make a few remarks on more recent applications of spheroidal functions, on extensive tables which are now in existence, and on a natural and useful generalization of the spheroidal functions.

As an appendix we attach a list of corrections of errors for the book by Meixner and Schäfke and close with a bibliography of relevant publications which have appeared since then, also emphasizing applications.

Finally, the authors share in this volume should be recorded. Their contributions to the theory in general can be inferred from the bibliography. With respect to this presentation there was a first German manuscript, sections 2.1.1. and 2.2. of which originate from G.Wolf, while almost all other contributions are due to F.W.Schäfke. On this basis the English manuscript was prepared by J.Meixner who also contributed the manuscript of 3.4. and compiled the bibliography. Of course, there were multifarious mutual stimulations, criticism and control.

1. Foundations.

1.1. Eigenvalue Problems with Two Parameters.

1.1.0. Introduction.

The theory of eigenvalue problems with two parameters, as presented in the following, starts from linear mappings

$$F,G,H,S : \mathfrak{U} \longrightarrow \mathfrak{R} ,$$

$$F^*,G^*,H^*,S^* : \mathfrak{U}^* \longrightarrow \mathfrak{R}^* \quad .$$

For $\mathfrak{R} \times \mathfrak{R}^*$ a bilinear "scalar product" $[\cdot,\cdot]$ is assumed. With respect to it there shall be "adjointness" in the form

$$[Fu,S^*v] = [Su,F^*v] \qquad (u \in \mathfrak{U}, v \in \mathfrak{U}^*) ,$$

the same for G,H in place of F. Then the two "adjoint" eigenvalue problems

$$Fy + \lambda Hy + \mu Gy = 0 \qquad (y \in \mathfrak{U}) ,$$

$$F^*y^* + \lambda H^*y^* + \mu G^*y^* = 0 \qquad (y^* \in \mathfrak{U}^*)$$

are considered. We regard μ as a perturbation parameter: the behavior of the eigenvalues λ, of eigen- and principal solutions in their dependence on μ and in comparison to the special case $\mu = 0$ are studied. The main intentions are power series expansions and asymptotic statements and finally equiconvergence theorems, which lead to expansion theorems for $\mu \neq 0$.

The basic presuppositions are formulated in a very general manner and most flexibly for multifarious applications. This concerns first the unperturbed problem and the three used generalized norms, $|\ |, \|\ \|, \sqcap$, in $\mathfrak{R}$, which leave much latitude in their properties and relations. Furthermore, the structure of the perturbation G is kept quite general; in contrast to MS, boundedness in any form is not more required. So, for boundary-eigenvalue problems also the treatment of perturbations of the boundary conditions becomes possible. Holomorphy, asymptotic statements and equiconvergence are formulated largely for terms of the form

$$[Bf,S^*y_n^*(\mu)]\ AHy_n(\mu)$$

with very general and expedient assumptions on the operators A,B, and again a wide range of applicability is obtained. Finally it is remarked that, as against the considerations in MS, the assumption of an entire function $y(\lambda,\mu)$, which for eigenvalue pairs gives the eigenfunctions, is renounced.

Apart from the much increased generality and adaptability, the realization of this program yields also in particular improvements in the technique of the proofs and sharper estimates.

Contrary to the approach in MS, which differs for the real case (I., II.) and the complex case (III.), the generalized theory permits the immediate application to both cases. In the complex case, it has the additional advantage that no consideration of sequences of norms or spaces, respectively, nor of (F)-spaces is necessary.

The theory includes throughout the possibility that for $\mu \neq 0$ multiple eigenvalues occur with corresponding systems of principal solutions. Thereby substantial gaps are closed as against earlier investigations. While in MS the expansion theorems in terms of Mathieu- and spheroidal functions are obtained only for "normal" values of h^2 and γ^2, respectively, here the "exceptional values" are comprised at once. For Hill's differential equation $y'' + (\lambda + \mu\varphi(x))y = 0$, the case of a non-even function φ and an integer characteristic exponent is at once included; here, for instance, in the study of expansions in terms of Bessel-functions, the treatment by a limiting procedure is no longer necessary or appropriate.

Summarizing: the theory which is developed in the following presents an attempt to cover in a most general and adaptable and nevertheless uniform way what seems to be accessible to the methodology formerly developed in MS.

Details of the disposition:

In 1.1.1. a first group of presuppositions is noted and expounded. More space is given to the assumptions on the used pseudonorms in (4) and to the formulation of the assumptions for the perturbation operator in (5). For that and for later use two, respectively four, types of operators are explained due to the possibility of certain estimates with the introduced norms; in doing so individual sequences of numbers in $\mathbb{R}^+$ play a part. These presuppositions suffice for the derivation of resolvent estimates in 1.1.2., which are the basis of all that follows. In particular, in 1.1.3. the desired statements on the eigenvalues λ to μ are obtained. After that, further presuppositions must be made in 1.1.4.. Thereby a concept of "order" for parameter - holomorphic operators and operator sequences, respectively, is employed, a concept which needs some attention. This permits to study the residues of the resolvent and to clarify its structure with appropriate principal solutions. Then a central result, a general equiconvergence theorem can be formulated and readily proved in 1.1.6.. In 1.1.7. considerations on (λ,μ) - and μ - holomorphy follow. 1.1.8. contains supplementary estimates which, on the one hand, lead to the asymptotics for large eigenvalue numbers, on the other hand, find application in the formulation of expansion theorems for $\mu \neq 0$ besides equiconvergence. 1.1.9. leads already over to the technique of the applications. It will be shown that for boundary-eigenvalue problems with ordinary differential equations in the real or in the complex domain the previously made presuppositions (2) and (7) are largely satisfied. The obtained results can be generalized and are of interest beyond those applications which are discussed here. Finally in 1.1.10. to 1.1.13. the application to Hill's differential equation (which includes

Mathieu's) and to the spheroidal differential equation in the real as well as in the complex domain are discussed.

We remark quite explicitly that the applications, which are discussed in this book, do by far not exhaust the consequences of the presented theory. Within the given frame and motive it was, however, necessary to forego numerous further considerations; they are reserved for publication elsewhere.

1.1.1. First presuppositions. Preliminary remarks.

We base our considerations on the assumptions which are put down and explained in the following.

(0) Let $\mathcal{R},\mathcal{R}^*,\mathcal{U},\mathcal{U}^*$ be linear spaces over $\mathbb{C}$. Let F,G,H,S be linear mappings of $\mathcal{U}$ into $\mathcal{R}$, F^*,G^*,H^*,S^* be linear mappings of $\mathcal{U}^*$ into $\mathcal{R}^*$. $[\ ,\ ]$ shall denote a bilinear mapping of $\mathcal{R}\times\mathcal{R}^*$ into $\mathbb{C}$. With respect to it, it is assumed that for $(u,v)\in\mathcal{U}\times\mathcal{U}^*$ and $(c_1,c_2,c_3)\in\mathbb{C}^3$ there holds

$$[(c_1F+c_2G+c_3H)u,\ S^*v]=[Su,(c_1F^*+c_2G^*+c_3H^*)v]\quad .$$

We consider now the eigenvalue problem

$$\text{(EP)}\qquad Fy+\mu Gy+\lambda Hy=0 \qquad\qquad (0\neq y\in\mathcal{U})$$

and the adjoint eigenvalue problem

$$\text{(aEP)}\qquad F^*y^*+\mu G^*y^*+\lambda H^*y^*=0 \qquad\qquad (0\neq y^*\in\mathcal{U}^*)\ .$$

In this connection we speak of an eigenvalue pair, for short: Evp, $(\lambda,\mu)\in\mathbb{C}^2$, or of an eigenvalue, for short: Ev, $\lambda\in\mathbb{C}$ to a fixed $\mu\in\mathbb{C}$.

(1) $Hy\neq 0$ for every eigensolution of (EP).

This means that for every $\mu\in\mathbb{C}$ the one-parametric eigenvalue problem is not degenerate: from $Hu=0$ and $Fu+\mu Gu=0$ it always follows that $u=0$.

(2) Let

$$\Delta:\mathbb{C}^2\to\mathbb{C}$$

be an entire analytic function. Assume that $\Delta(\lambda,\mu)=0$ if and only if (λ,μ) is Evp of (EP) and if and only if (λ,μ) is Evp of (aEP).

(3) To $\mu=0$ there shall be precisely an infinite countable set of eigenvalues λ_n $(n\in\mathbb{N})$ which are counted according to their multiplicity as zeros of $\Delta(\cdot,0)$. For $n\in\mathbb{N}$ the y_n^* shall be solutions of (aEP) to $(\lambda_n,0)$. If $\lambda_{n_1}=\lambda_{n_2}=\dots=\lambda_{n_k}$ with different subscripts, then the $y_{n_1}^*,y_{n_2}^*,\dots,y_{n_k}^*$ shall be linearly independent.

In the following we shall use the concept "pseudonorm". It has the properties of a norm in a linear space except for definiteness and finiteness.

(4) In the space $\mathbb{C}^{\mathbb{N}}$ of all sequences of complex numbers $\alpha = (\alpha_n)_{n \in \mathbb{N}}$ let
$|\alpha|_\infty := \sup|\alpha_n|$, $|\alpha|_1 := \sum|\alpha_n|$, $|\alpha|_2 := \left(\sum|\alpha_n|^2\right)^{1/2}$
be the usual totally subadditve definite pseudonorms. Moreover, let $\|\ \|_\infty$, $\|\ \|_2$, $\|\ \|_1$ be totally subadditive definite pseudonorms in $\mathbb{C}^{\mathbb{N}}$ which are related to the first ones via fixed sequences

$$\delta = (\delta_n)_{n \in \mathbb{N}}, \quad \delta_n > 0 \qquad (n \in \mathbb{N}),$$

$$\eta = (\eta_n)_{n \in \mathbb{N}}, \quad \eta_n > 0 \qquad (n \in \mathbb{N})$$

by

(4.1) $$\|\alpha\cdot\beta\|_2 \leqq |\alpha|_\infty \cdot \|\beta\|_2 \quad ,$$

(4.2) $$\|\alpha\cdot\beta\|_2 \leqq |\alpha\cdot\delta|_2 \, \|\beta\|_\infty \quad ,$$

(4.3) $$\|\alpha\cdot\beta\|_1 \leqq |\alpha\cdot\eta|_2 \, \|\beta\|_2 \quad .$$

The linear mapping

$$\varphi : \mathcal{R} \longrightarrow \mathbb{C}^{\mathbb{N}}$$

is defined by

(4.4) $$\varphi(f) := ([f, S^* y_n^*])_{n \in \mathbb{N}} \; .$$

Let $\|\ \|, \ulcorner\!\urcorner, |\ |_o$ be pseudonorms in $\mathcal{R}$ which are related to the preceding ones by

(4.5) $$\|f\| = \|\varphi(f)\|_2 \; ,$$

(4.6) $$\ulcorner f \urcorner \geqq \|\varphi(f)\|_\infty \; ,$$

(4.7) $$|f|_o \leqq \|\varphi(f)\|_1 \; .$$

$\|\ \|$ shall be definite in $H\mathcal{U}$.

In the following definitions A shall be a linear mapping of a subspace ϑ_A of $\mathcal{R}$ onto a subspace $\mathcal{R}_A$ of $\mathcal{R}$ and $\alpha = (\alpha_n)_{n \in \mathbb{N}}$ a sequence of non-negative numbers.

We say, provided that $\vartheta_A \supset H\mathcal{U}$:

$$A \in (I,1;\alpha) \quad \text{if} \quad \|Af\| \leqq \|\alpha\cdot\varphi(f)\|_2 \qquad (f \in \vartheta_A)$$

and

$$A \in (I,2;\alpha) \quad \text{if} \quad |Af|_o \leqq \|\alpha\cdot\varphi(f)\|_1 \qquad (f \in \vartheta_A) \; .$$

We say:

$$A \in (II,1;\alpha) \quad \text{if} \begin{cases} |\varphi(Af)| \leqq \alpha\cdot\psi(f) \; , \\ \|\psi(f)\|_2 \leqq \|f\| \; , \\ \psi : \vartheta_A \longrightarrow \mathbb{R}_+^{\mathbb{N}} \; , \end{cases} \qquad (f \in \vartheta_A) \; .$$

and

$$A \in (II,2;\alpha) \quad \text{if} \begin{cases} |\varphi(Af) \leqq \alpha \cdot \psi(f) , \\ \|\psi(f)\|_\infty \leqq \lceil f \rceil , \\ \psi : \vartheta_A \longrightarrow \mathbb{R}_+^{\mathbb{N}} . \end{cases} \qquad (f \in \vartheta_A)$$

With a view to applications we note two lemmas.

Lemma I: Let for $k \in \mathbb{N}$

$$D_k : \vartheta_A \longrightarrow \mathcal{U}_k \subset \mathcal{R} \text{ linear,}$$
$$C_k : \mathcal{U}_k \longrightarrow \mathcal{R} \text{ linear}$$

and $\varepsilon_k \geqq 0$ with $\sum \varepsilon_k \leqq 1$ such that termwise

(1) $\varphi(Af) = \sum \varphi(C_k D_k f)$ $\qquad (f \in \vartheta_A)$,

(2) $|\varphi(D_k f)| \leqq \alpha|\varphi(f)|$ $\qquad (f \in \vartheta_A)$,

(3) $\|C_k g\| \leqq \varepsilon_k \|g\|$ $\qquad (g \in \mathcal{U}_k)$.

Then $A \in (I,1;\alpha)$.

If (3) is replaced by

(3') $\|\varphi(C_k g)\|_1 \leqq \varepsilon_k \|\varphi(g)\|_1$ $\qquad (g \in \mathcal{U}_k)$,

then $A \in (I,2;\alpha)$.

Proof: For instance the second statement is obtained by virtue of (4.7),(1), (3') and (2) from

$$|Af|_o \leqq |\varphi(Af)\|_1 \leqq \sum \|\varphi(C_k D_k f)\|_1$$
$$\leqq \sum \varepsilon_k \|\varphi(D_k f)\|_1 \leqq \|\alpha \cdot \varphi(f)\|_1 .$$

Lemma II: Let for $k \in \mathbb{N}$

$$C_k : \vartheta_A \longrightarrow \mathcal{U}_k \subset \mathcal{R} \text{ linear,}$$
$$D_k : \mathcal{U}_k \longrightarrow \mathcal{R} \text{ linear}$$

and $\varepsilon_k \geqq 0$ with $\sum \varepsilon_k \leqq 1$ such that termwise

(1) $\varphi(Af) = \sum \varphi(D_k C_k f)$ $\qquad (f \in \vartheta_A)$,

(2) $|\varphi(D_k g)| \leqq \alpha \cdot |\varphi(g)|$ $\qquad (g \in \mathcal{U}_k)$,

(3) $\|C_k f\| \leqq \varepsilon_k \|f\|$ $\qquad (f \in \vartheta_A)$.

Then $A \in (II,1;\alpha)$.

If (3) is replaced by

(3') $\lceil C_k f \rceil \leqq \varepsilon_k \lceil f \rceil$ $\qquad (f \in \vartheta_A)$,

then $A \in (II,2;\alpha)$.

Proof: For instance, for the second statement one starts with

$$|\varphi(Af)| \leqq \sum |\varphi(D_k C_k f)| \leqq \alpha \cdot \sum |\varphi(C_k f)|$$

to obtain

$$|\varphi(Af)| \leq \alpha \cdot \psi(f) \qquad (f \in \vartheta_A)$$

with

$$\psi(f) := \sum |\varphi(C_k f)| \ .$$

Then one proceeds as follows:

$$\|\psi(f)\|_\infty \leq \sum \|\varphi(C_k f)\|_\infty \leq \sum \lceil C_k f \rceil \leq \sum \varepsilon_k \lceil f \rceil \leq \lceil f \rceil \ .$$

Remark: In the Lemmata I und II the inequality (2) certainly holds if for $(k,n) \in \mathbb{N}^2$, and $f \in \vartheta_A$ or $f \in \mathcal{U}_k$, respectively, there holds

$$[D_k f, S^* y_n^*] = \mu_{nk}[f, S^* y_n^*], \ |\mu_{nk}| \leq \alpha_n \ .$$

Now we require for G

(5) a) Let for $k \in \mathbb{N}$

$A_k \in (I,1;\gamma^{(1)})$, $B_k \in (II,1;\gamma^{(2)})$

with fixed sequences $\gamma^{(1)}, \gamma^{(2)}$ and $\mathcal{R}_{A_k} \subset \vartheta_{B_k}$.

With these there shall hold for $f \in \mathcal{U}$ (termwise)

$$\varphi(Gf) = \sum_{k \in \mathbb{N}} \zeta_k \varphi(B_k A_k Hf), \ \sum_{k \in \mathbb{N}} |\zeta_k| \leq 1 \ .$$

b) With

$$\gamma := \gamma^{(1)} \cdot \gamma^{(2)} \ ,$$

there shall be

$$\gamma_{n_1} = \gamma_{n_2} \quad \text{if} \quad \lambda_{n_1} = \lambda_{n_2}$$

and

$$\gamma_n = \mathcal{O}(\lambda_n) \qquad (n \to \infty) \ .$$

c) If y is an eigensolution of (EP) to an Evp (λ,μ) with $\lambda \neq \lambda_n$, then there shall be

$$\sup_{k \in \mathbb{N}} \|A_k H y\| < \infty \ .$$

A sufficient condition for (5) c) to hold will be given in <u>1.1.2.</u> . (5) b) entails that for $\lambda \neq \lambda_n$

(5.1) $$M_\gamma(\lambda) := \sup_{n \in \mathbb{N}} \gamma_n |\lambda - \lambda_n|^{-1} < \infty \ .$$

Then we set

(5.2) $$m_\gamma(\lambda) := \begin{cases} 0 & (\lambda = \lambda_n, \ n \in \mathbb{N}) \\ M_\gamma(\lambda)^{-1} & (\text{else}) \end{cases} \ ,$$

with $0^{-1} := +\infty$.

We also define

(5.3) $$\hat{M} := \sup\{m_\gamma(\lambda) : \lambda \in \mathbb{C}\} \ .$$

Finally we assume

(6) If (λ,μ) is not an Evp, then for $f \in \mathcal{R}$

$$Fz + \lambda Hz + \mu Gz = f \qquad (z \in \mathcal{U})$$

shall have a (unique) solution.

Then we designate

$$z =: R(\lambda,\mu)f$$

and remark that this "resolvent" represents a linear mapping

$$R(\lambda,\mu) : \mathcal{R} \longrightarrow \mathcal{U}.$$

1.1.2. Estimates for the Resolvent.

At first we consider estimates with $R(\lambda,0)$ and $\lambda \neq \lambda_n$ $(n \in \mathbb{N})$. We start from

$$Fz + \lambda Hz = f \; ,$$

$$F^*y_n^* + \lambda_n H^*y_n^* = 0 \; ,$$

apply $[\cdot, S^*y_n^*]$ to the first, $[Sz,\cdot]$ to the second equation and take the difference of the results. Subject to the adjointness required in (0) we obtain

$$[Hz, S^*y_n^*] = \frac{1}{\lambda-\lambda_n} [f, S^*y_n^*].$$

With the notation

(*)
$$\ell(\lambda) := \left(\frac{1}{\lambda-\lambda_n}\right)_{n\in\mathbb{N}} \quad \text{for } \lambda \neq \lambda_n \quad (n \in \mathbb{N}) \; ,$$

this can, according to (4), also be written as

(x)
$$\varphi(HR(\lambda,0)f) = \ell(\lambda) \cdot \varphi(f) \; .$$

In the following we call a sequence $\alpha = (\alpha_n)_{n\in\mathbb{N}}$ with $\alpha_n \geqq 0$ "relatively bounded" if

$$\alpha_n = \mathcal{O}(\lambda_n) \qquad (n \to \infty)$$

and define for $\lambda \neq \lambda_n$

$$M_\alpha(\lambda) := \sup_{n\in\mathbb{N}} \frac{\alpha_n}{|\lambda-\lambda_n|} = |\alpha \cdot \ell(\lambda)|_\infty \; .$$

We designate α as "subdominant" if, moreover,

$$\sum_{\lambda_n \neq 0} \alpha_n^2 |\lambda_n|^{-2} < \infty \; ;$$

in this case we define for $\lambda \neq \lambda_n$

$$\sigma_\alpha(\lambda) := \Big(\sum_{n\in\mathbb{N}} \alpha_n^2 |\lambda-\lambda_n|^{-2}\Big)^{1/2} = |\alpha \cdot \ell(\lambda)|_2 \; .$$

Now we are in a position to formulate

<u>Theorem 1:</u> Let $\lambda \neq \lambda_n$ $(n \in \mathbb{N})$.

If $A \in (I,1;\alpha)$, $B \in (II,1;\beta)$, $f \in \vartheta_B$ and if $\alpha \cdot \beta$ is relatively bounded, then

(1.1) $$\|AHR(\lambda,0)Bf\| \leqq M_{\alpha\beta}(\lambda)\|f\| .$$

If $A \in (I,1;\alpha)$, $B \in (II,2;\beta)$, $f \in \vartheta_B$ and if $\alpha\beta\delta$ is subdominant, then

(1.2) $$\|AHR(\lambda,0)Bf\| \leqq \sigma_{\alpha\beta\delta}(\lambda)\lceil f \rceil .$$

If $A \in (I,2;\alpha)$, $B \in (II,1;\beta)$, $f \in \vartheta_B$ and if $\alpha\beta\eta$ is subdominant, then

(1.3) $$|AHR(\lambda,0)Bf|_o \leqq \sigma_{\alpha\beta\eta}(\lambda)\ \|f\| .$$

<u>Proof:</u> (1.1) results from

$$\|AHR(\lambda,0)Bf\| \leqq \|\alpha\cdot \varphi(HR(\lambda,0)Bf)\|_2 \leqq \|\alpha\cdot \ell(\lambda) \cdot \varphi(Bf)\|_2 \leqq$$
$$\leqq \|\alpha\cdot\beta\cdot\ell(\lambda)\cdot\psi(f)\|_2 \leqq M_{\alpha\beta}(\lambda)\|\psi(f)\|_2 \leqq M_{\alpha\beta}(\lambda)\|f\| .$$

Here, in turn, $A \in (I,1;\alpha)$, (×), $B \in (II,1;\beta)$ with the monotony of $\|\ \|_2$, (4.1), and again $B \in (II,1;\beta)$, have been used. - For (1.2) the first four steps are the same. But then one proceeds by

$$\|\alpha\cdot\beta\cdot\ell(\lambda) \cdot \psi(f)\|_2 \leqq \sigma_{\alpha\beta\delta}(\lambda) \cdot\| \psi(f)\|_\infty \leqq \sigma_{\alpha\beta\delta}(\lambda)\lceil f \rceil$$

with use of (4.2) and $B \in (II,2;\beta)$. - (1.3) results from

$$|AHR(\lambda,0)Bf|_o \leqq \|\ \alpha\cdot \varphi\ (HR(\lambda,0)Bf)\|_1 \leqq \|\ \alpha\cdot \ell\ (\lambda) \cdot \varphi(Bf)\|_1 \leqq$$
$$\leqq \|\alpha \cdot \beta \cdot \ell(\lambda) \cdot \psi(f)\|_1 \leqq \sigma_{\alpha\beta\eta}(\lambda) \cdot\|\psi\ (f)\|_2 \leqq \sigma_{\alpha\beta\eta}(\lambda)\ \|f\| ,$$

where, in turn, $A \in (I,2;\alpha)$, (×), $B \in (II,1;\beta)$ with the monotony of $\|\ \|_1$, (4.3), and again $B \in (II,1;\beta)$, have been used.

Now we can prove

<u>Theorem 2:</u> For an eigenvalue pair (λ,μ) there holds
$$m_\gamma(\lambda) \leqq |\mu| .$$
Therefore for $|\mu| < \hat{M}$ there is $\Delta(\cdot,\mu) \neq 0$.

<u>Proof:</u> According to (5.2) it suffices to assume $\lambda \neq \lambda_n$ $(n \in \mathbb{N})$. Then one rewrites (EP) as
$$y = -\mu R(\lambda,0)Gy$$
and obtains for $k \in \mathbb{N}$ according to ⑤
$$A_k Hy = -\mu A_k HR(\lambda,0)Gy .$$
Here one estimates as for (1.1) in Theorem 1:
$$\begin{aligned}\|A_k Hy\| &\leqq |\mu|\ \|\gamma^{(1)} \cdot \ell(\lambda) \cdot \varphi(Gy)\|_2 \leqq \\ &\leqq |\mu| \sup_{j\in\mathbb{N}}\|\gamma^{(1)} \cdot \ell(\lambda) \cdot \varphi(B_j A_j Hy)\|_2 \leqq \\ &\leqq |\mu| \sup_{j\in\mathbb{N}}\|\gamma^{(1)} \cdot |\ell(\lambda)| \cdot \gamma^{(2)} \cdot \psi_j(A_j Hy)\|_2 \leqq \\ &\leqq |\mu|\ M_\gamma(\lambda) \sup_{j\in\mathbb{N}} \|A_j Hy\| .\end{aligned}$$

Now

$$0 < \sup_{j \in \mathbb{N}} \|A_j Hy\| < \infty$$

where the right inequality follows from (5) c). In the left inequality the equality sign must be excluded because it would entail $\|A_j Hy\| = 0 \quad (j \in \mathbb{N})$ and with $B_j \in (II,1;\gamma^{(2)})$ one would obtain $\psi_j(A_j Hy) = 0$, $\varphi(B_j A_j Hy) = 0$, therefore $\varphi(Gy) = 0$ and $\|Gy\| = 0$.

Now

$$Hy = -\mu HR(\lambda,0)Gy$$

and (1.1) of Theorem 1 with $A = id_{H\mathcal{U}}$, $B = id_{\mathcal{R}}$, $\alpha = \beta = 1 := (1)$ can be applied. Since 1 is obviously relatively bounded, one would obtain

$$\|Hy\| \leqq |\mu| M_1(\lambda) \|Gy\| \quad ,$$

and there would result $\|Hy\| = 0$. Together with the last requirement of (4) this would yield $Hy = 0$ in contradiction to (1).

Therefore our estimates lead to

$$1 \leqq |\mu| \; M_\gamma(\lambda)$$

which is the assertion due to (5.2). -

We complete now the

Remark to (5) c) : For (5) c) to hold it is sufficient that $\gamma \cdot \delta$ is subdominant, $B_k \in (II,2;\gamma^{(1)})$ and that $\sup_{j \in \mathbb{N}} \overline{|A_j Hy|} < \infty$ for the y characterized in (5) c).

Indeed one estimates as in the proof of Theorem 2

$$\|A_k Hy\| \leqq |\mu| \sup_{j \in \mathbb{N}} \| \gamma \cdot |\ell(\lambda)| \cdot \psi_j(A_j Hy)\|_2$$

and obtains, furthermore, with (4.2), (4.6)

$$\leqq |\mu| \sigma_{\gamma\delta}(\lambda) \sup_{j \in \mathbb{N}} \overline{|A_j Hy|}$$

from which the validity of the Remark is recognized.

We get now to estimates for $R(\lambda,\mu)$ and formulate

Theorem 3 : Let $m_\gamma(\lambda) > |\mu|$.

If $A \in (I,1;\alpha)$, $B \in (II,1;\beta)$ and if $\alpha \cdot \gamma^{(2)}$, $\beta \cdot \gamma^{(1)}$ are relatively bounded, then for $f \in \mathcal{D}_B$

$$(3.1) \qquad \|AH(R(\lambda,\mu) - R(\lambda,0))Bf\| \leqq \frac{|\mu| M_{\alpha\gamma^{(2)}}(\lambda) M_{\beta\gamma^{(1)}}(\lambda)}{1 - |\mu| M_\gamma(\lambda)} \|f\| \quad .$$

If $A \in (I,1;\alpha)$, $B \in (II,2;\beta)$ and if $\alpha \cdot \gamma^{(2)}$ is relatively bounded and $\beta \cdot \gamma^{(1)} \cdot \delta$ is subdominant, then

$$(3.2) \qquad \|AH(R(\lambda,\mu) - R(\lambda,0))Bf\| \leqq \frac{|\mu| M_{\alpha\gamma^{(2)}}(\lambda) \sigma_{\beta\gamma^{(1)}\delta}(\lambda)}{1 - |\mu| M_\gamma(\lambda)} \overline{|f|} \quad .$$

If $A \in (I,2;\alpha)$, $B \in (II,1;\beta)$ and if $\alpha \cdot \gamma^{(2)} \cdot \eta$ is subdominant and $\beta \cdot \gamma^{(1)}$

is relatively bounded, then

(3.3) $$|AH(R(\lambda,\mu) - R(\lambda,0))Bf|_0 \leq \frac{|\mu|\sigma_{\alpha\gamma^{(2)}\eta}(\lambda)M_{\beta\gamma^{(1)}}(\lambda)}{1 - |\mu|M_\gamma(\lambda)} \|f\| .$$

If $A \in (I,2;\alpha)$, $B \in (II,2;\beta)$ and if $\alpha \cdot \gamma^{(2)} \cdot \eta$ and $\beta \cdot \gamma^{(1)} \cdot \delta$ are subdominant, then

(3.4) $$|AH(R(\lambda,\mu) - R(\lambda,0))Bf|_0 \leq \frac{|\mu|\sigma_{\alpha\gamma^{(2)}\eta}(\lambda)\sigma_{\beta\gamma^{(1)}\delta}(\lambda)}{1 - |\mu|M_\gamma(\lambda)} \lceil f \rceil .$$

<u>Proof:</u> After Theorem 2 (λ,μ) and $(\lambda,0)$ are not eigenvalue pairs; therefore $R(\lambda,\mu)$ and $R(\lambda,0)$ are defined by (6) ; according to <u>1.</u>, (5.2) one has, moreover,

$$|\mu|M_\gamma(\lambda) < 1 .$$

If one rewrites

$$Fz + \mu Gz + \lambda Hz = g , \quad z = R(\lambda,\mu)g$$

as

$$Fz + \lambda Hz = g - \mu Gz ,$$

one has the "resolvent equation"

$$R(\lambda,\mu)g = R(\lambda,0)g - \mu R(\lambda,0)GR(\lambda,\mu)g .$$

At first one considers according to (5)

$$A_k HR(\lambda,\mu)g = A_k HR(\lambda,0)g - \mu A_k HR(\lambda,0)GR(\lambda,\mu)g$$

and can estimate as in the proof of Theorem 2

$$\|A_k HR(\lambda,\mu)g\| \leq \|A_k HR(\lambda,0)g\| + |\mu|M_\gamma(\lambda) \sup_{j\in\mathbb{N}} \|A_j HR(\lambda,\mu)g\| .$$

This leads to

(+) $$\sup_{k\in\mathbb{N}} \|A_k HR(\lambda,\mu)g\| \leq (1 - |\mu|M_\gamma(\lambda))^{-1} \sup_{k\in\mathbb{N}} \|A_k HR(\lambda,0)g\| .$$

Now one writes, using the resolvent equation,

$$AH(R(\lambda,\mu) - R(\lambda,0))Bf = -\mu\, AHR(\lambda,0)GR(\lambda,\mu)Bf$$

and estimates the left member. If $A \in (I,1;\alpha)$ and if $\alpha \cdot \gamma^{(2)}$ is relatively bounded, then one obtains as in the proof of Theorem 2

$$\|AHR(\lambda,0)GR(\lambda,\mu)Bf\| \leq M_{\alpha\gamma^{(2)}}(\lambda) \sup_{k\in\mathbb{N}} \|A_k HR(\lambda,\mu)Bf\| .$$

This is combined with (+) and Theorem 1, (1.1) and (1.2), respectively, are applied. Then the statements (3.1) and (3.2) are obtained. If on the other hand $A \in (I,2;\alpha)$ and $\alpha \cdot \gamma^{(2)} \cdot \eta$ is subdominant, then one obtains as for (1.3)

$$|AHR(\lambda,0)Gz|_0 \leq \|\alpha\rho(HR(\lambda,0)Gz)\|_1 \leq \|\alpha \cdot \ell(\lambda) \cdot \varphi(Gz)\|_1 \leq$$

$$\leq \sup_{k\in\mathbb{N}} \|\alpha \cdot \gamma^{(2)} \cdot \ell(\lambda) \cdot \psi_k(A_k Hz)\|_1 \leq$$

$$\leq \sigma_{\alpha\gamma^{(2)}\eta}(\lambda) \sup_{k\in\mathbb{N}} \|\psi_k(A_k Hz\|_2 \leq$$

$$\leq \sigma_{\alpha\gamma^{(2)}\eta}(\lambda) \sup_{k \in \mathbb{N}} \|A_k Hz\| \quad .$$

One introduces $z = R(\lambda,\mu)Bf$, combines here again with (+) and applies (1.1) and (1.2), respectively, of Theorem 1. Then one obtains just the statements (3.3) and (3.4).

From the formulas of the proof one can infer for later use for instance

Theorem 3a: If $A \in (I,2;\alpha)$, $B \in (II,1;\beta)$, if $\alpha\gamma^{(2)}\eta$ is subdominant and $\beta\gamma^{(1)}$ is relatively bounded, then for $m_\gamma(\lambda) > |\mu|$, $m_\gamma(\lambda_o) > |\mu_o|$, $f \in \mathcal{D}_B$, there hold

$$|AHR(\lambda,\mu)GR(\lambda_o,\mu_o)Bf|_o \leq \frac{\sigma_{\alpha\gamma^{(2)}\eta}(\lambda)M_{\beta\gamma^{(1)}}(\lambda_o)}{(1-|\mu|M_\gamma(\lambda))(1-|\mu_o|M_\gamma(\lambda_o))} \|f\| \quad .$$

and

$$|AHR(\lambda,\mu)GR(\lambda,\mu)GR(\lambda_o,\mu_o)Bf|_o \leq \frac{\sigma_{\alpha\gamma^{(2)}\eta}(\lambda)M_\gamma(\lambda)M_{\beta\gamma^{(1)}}(\lambda_o)}{(1-|\mu|M_\gamma(\lambda))^2(1-|\mu_o|M_\gamma(\lambda_o))} \|f\| \quad .$$

If $A \in (I,1;\alpha)$, if $\alpha\gamma^{(2)}$ is relatively bounded, else as above, there holds

$$\|AHR(\lambda,\mu)GR(\lambda_o,\mu_o)Bf\| \leq \frac{M_{\alpha\gamma^{(2)}}(\lambda)M_{\beta\gamma^{(1)}}(\lambda_o)}{(1-|\mu|M_\gamma(\lambda))(1-|\mu_o|M_\gamma(\lambda_o))} \|f\| \quad .$$

Proof: From

$$AHR(\lambda,\mu)Gz = AHR(\lambda,0)Gz - \mu AHR(\lambda,0)GR(\lambda,\mu)Gz$$

it follows with the last estimate of the proof to Theorem 3 that

$$|AHR(\lambda,\mu)Gz|_o \leq \sigma_{\alpha\gamma^{(2)}\eta}(\lambda)[\sup\|A_k Hz\| + |\mu| \cdot \sup\|A_k HR(\lambda,\mu)Gz\|] \quad .$$

For the last term one uses (+) in the above proof and estimates then as for Theorem 2. This yields

$$|AHR(\lambda,\mu)Gz|_o \leq \sigma_{\alpha\gamma^{(2)}\eta}(\lambda)(1-|\mu|M_\gamma(\lambda))^{-1} \sup\|A_k Hz\| \quad .$$

Now $z = R(\lambda_o,\mu_o)Bf$ is introduced and (+) together with Theorem 1 are used. This gives the first assertion. The second one follows by analogy by means of iteration. The third one also follows by analogy.

1.1.3. The Eigenvalues to $\mu \neq 0$.

The following considerations in regard of the eigenvalues λ to $\mu \neq 0$ are based essentially on (2) and Theorem 3.

Of importance in this connection are (with $0^{-1} := \infty$)

$$0 < \hat{L} := \liminf_{n\to\infty} \gamma_n^{-1}|\lambda_n| \leq \infty$$

and for c with $0 \leq c \leq \infty$ the sets

$$\mathfrak{M}_c := \{\lambda \in \mathbb{C} : m_\gamma(\lambda) \leq c\} ,$$

which are obviously closed.

In this connection some properties of m_γ, defined by 1.1.1., (5.1), (5.2), must be noticed. They require somewhat more attention because of the possibilities $\gamma = 0$ or $\gamma_m = 0$ as against the simple case that all $\gamma_n > 0$.

At first, we have in every case $m_\gamma(\lambda) = 0$ if and only if $\lambda = \lambda_{n_0}$, $n_0 \in \mathbb{N}$. If $\gamma = 0$, then $m_\gamma(\lambda) = +\infty$ for $\lambda \in \mathbb{C} \setminus \{\lambda_n : n \in \mathbb{N}\}$. If $\gamma \neq 0$, then there is always $m_\gamma(\lambda) < \infty$; in this case m_γ is continuous at all points except for the λ_m with $\gamma_m = 0$; at these points there exists

$$\lim_{\lambda \ddagger \lambda_m} m_\gamma(\lambda) = M_\gamma(\lambda_m)^{-1} > 0$$

with

$$M_\gamma(\lambda_m) := \lim_{\lambda \ddagger \lambda_m} M_\gamma(\lambda) = \sup_{\lambda_n \neq \lambda_m} \frac{\gamma_n}{|\lambda_n - \lambda_m|} .$$

Now we note at first

Theorem 4 : a) There holds (see 1.1.1. (5.3))

$$\hat{M} \leq \hat{L} .$$

b) There is $\mathfrak{M}_c = \mathbb{C}$ if and only if $c \geq \hat{M}$.

c) If $0 < m_\gamma(\lambda) < \hat{L}$, then
$m_\gamma(\lambda) = \min \gamma_n^{-1} |\lambda - \lambda_n|$.

d) Consequently for $0 \leq c < \hat{M}$ there holds

$$\mathfrak{M}_c = \bigcup_{n \in \mathbb{N}} \{\lambda \in \mathbb{C} : |\lambda - \lambda_n| \leq \gamma_n c\} .$$

This follows almost immediately from the definitions together with $\lambda_n \to \infty$.

Now we obtain

Theorem 5 : Let $(0 \leq c < \hat{M}$ and) $\mathfrak{K}$ be a compact component of $\mathfrak{M}_c$.

Then $\mathfrak{K}$ is the union of a finite number of "circular disks" $\{\lambda \in \mathbb{C} : |\lambda - \lambda_n| \leq \gamma_n c\}$ For $|\mu| \leq c$ the number of eigenvalues λ in $\mathfrak{K}$, each one counted according to its order as zeros of $\Delta(\cdot,\mu)$, is constant. The symmetric polynomials of these eigenvalues are for $|\mu| \leq c$ holomorphic.

Proof : The first statement is obvious in consequence of Theorem 4,d) and because of $\lambda_n \to \infty$. Now there exists a closed, continuous and rectifiable contour $\mathcal{L}$ which runs once around $\mathfrak{K}$ in the positive sense but does not run around $\mathfrak{M}_c \setminus \mathfrak{K}$, and on which $m_\gamma(\lambda) \geq c' > c$. Then from Theorem 3 it follows that $\Delta(\lambda,\mu) \neq 0$ for $|\mu| < c'$ and λ on $\mathcal{L}$. Thus one can form for $|\mu| < c'$ and $\sigma = 0,1,2,\ldots$

$$\frac{1}{2\pi i} \int_{\mathcal{L}} \frac{\frac{\partial}{\partial \lambda} \Delta(\lambda,\mu)}{\Delta(\lambda,\mu)} \lambda^\sigma d\lambda$$

which is obviously holomorphic for $|\mu| < c'$. Since the zeros of $\Delta(\cdot,\mu)$ are in $\mathfrak{K}$ or in $\mathfrak{M}_c \setminus \mathfrak{K}$, and have the circuital numbers 1 and 0, respectively, one obtains for $\sigma = 0$ the consequently constant number of eigenvalues λ to μ in $\mathfrak{K}$, and for $\sigma \in \mathbb{N}$ the corresponding sums of powers of λ. This gives the assertion.

We infer from this proof immediately

Theorem 6 : Let $0 \leq c < \hat{M}$. $\mathfrak{M}_c$ divides into a countable number of compact components precisely if for $n \in \mathbb{N}$ there exist closed, continuous and rectifiable contours $\mathcal{L}_n$ on which $m_\gamma(\lambda) > c$ and which possess a circuital number $\neq 0$ around the respective λ_n.

In the following we shall consider the case that $\{\lambda \in \mathbb{C} : |\lambda - \lambda_m| \leq \gamma_m c\}$ is a compact component of $\mathfrak{M}_c$ which does not contain any $\lambda_n \neq \lambda_m$. For that purpose we define

$$r_m := \begin{cases} +\infty, \text{ if } \gamma = 0, \\ M_\gamma(\lambda_m)^{-1}, \quad \text{if } \gamma \neq 0, \gamma_m = 0, \\ \max\left\{ \frac{\rho}{\gamma_m} : m_\gamma(\lambda) = \frac{|\lambda-\lambda_m|}{\gamma_m} \ (|\lambda - \lambda_m| \leq \rho)\right\}, \text{ if } \gamma_m \neq 0. \end{cases}$$

Then there holds obviously

Theorem 7 : If and only if $0 \leq c < r_m$ is $\{\lambda \in \mathbb{C} : |\lambda - \lambda_m| \leq \gamma_m c\}$ a compact component of $\mathfrak{M}_c$ which contains no $\lambda_n \neq \lambda_m$.

We set now

$$d_m := \min\{|\lambda_n - \lambda_m| : \lambda_n \neq \lambda_m\}$$

and demonstrate

Theorem 8 : There hold

$$r_m \leq d_m \cdot \gamma_m^{-1},$$

$$\liminf_{n\to\infty} r_n \geq \frac{1}{2} \liminf_{n\to\infty} d_n \cdot \gamma_n^{-1}$$

(with $0^{-1} := +\infty$). Consequently $r_n \to \infty$ if and only if $d_n \cdot \gamma_n^{-1} \to \infty$.

Proof : The first statement is clear according to the definitions. Let for the proof of the second statement

$$0 < c < \frac{1}{2} \liminf_{n\to\infty} d_n \cdot \gamma_n^{-1}.$$

Then one has to show that $r_n < c$ can hold only a finite number of times. Let $r_n < c$, therefore $\gamma \neq 0$. Then in each case there exist a $\lambda' \in \mathbb{C}$ and an $m \in \mathbb{N}$ with

$$|\lambda' - \lambda_n| \leq \gamma_n c, \ |\lambda' - \lambda_m| \leq \gamma_m c, \ \lambda_m \neq \lambda_n.$$

Consequently

(*) $$0 < |\lambda_n - \lambda_m| \leq (\gamma_n + \gamma_m)c$$

and a fortiori

$$\frac{1}{2} \min\left(d_n \cdot \gamma_n^{-1}, d_m \cdot \gamma_m^{-1}\right) \leq c.$$

Now obviously

$$c < \hat{C}.$$

Therefore for fixed n in (*) only a finite number of values m is suitable, and conversely. Now: if $r_n < c$ would occur infinitely many times, this would mean that $\frac{1}{2} d_k \cdot \gamma_k^{-1} \leqq c$ would hold infinitely many times - in contradiction to the choice of c. - The last statement follows from the first two ones.

Theorem 7 and Theorem 8 yield

<u>Theorem 9 :</u> If

$$0 \leqq c < \liminf r_n$$

or even

$$0 \leqq c < \frac{1}{2} \liminf d_n \cdot \gamma_n^{-1} ,$$

then $\mathfrak{M}_c$ divides into a countable number of compact components, all of which, apart from a finite number, are of the "circular disk type" mentioned in Theorem 7.

Theorem 5 and Theorem 7 yield.

<u>Theorem 10 :</u> If λ_n is a simple zero of $\Delta(\cdot,0)$, then $\Delta(\cdot,\mu)$ has for $|\mu| < r_n$ precisely one zero $\lambda_n(\mu)$ with $|\lambda_n(\mu) - \lambda_n| \leqq \gamma_n|\mu|$. There exists a power series expansion

$$\lambda_n(\mu) = \lambda_n + \lambda_{n,1} \mu + \lambda_{n,2}\mu^2 + \dots$$

which converges at least for $|\mu| < r_n$ and the coefficients have estimates

$$|\lambda_{n,k}| \leqq \gamma_n \cdot r_n^{-(k-1)} \qquad (k \in \mathbb{N}) .$$

Therefrom there results together with Theorem 8

<u>Theorem 11 :</u> If all λ_n except for a finite number are simple zeros of $\Delta(\cdot,0)$ and if $d_n \cdot \gamma_n^{-1} \to \infty$, then the power series of Theorem 10 are at the same time asymptotic series for $n \to \infty$: one has in fact

$$\lambda_n(\mu) = \lambda_n + \lambda_{n1}\mu + \dots + \lambda_{nk}\mu^k + \mathcal{O}\left(\gamma_n r_n^{-k}\right) .$$

Moreover, we note

<u>Theorem 12 :</u> If all λ_n $(n \in \mathbb{N})$ are simple zeros of $\Delta(\cdot,0)$ and if $0 < c < \liminf r_n$, then for all μ with $|\mu| \leqq c$, except for a finite number of them, all the eigenvalues λ to μ are simple zeros of $\Delta(\cdot,\mu)$.

<u>Proof :</u> According to Theorem 9, Theorem 7, and Theorem 10 the $\lambda_n(\mu)$ with $|\mu| \leqq c$ and $r_n > c$ are simple zeros of $\Delta(\cdot,\mu)$. There remains a finite number of eigenvalues λ to μ, which according to Theorem 5 is equal to the number of the λ_n with $r_n \leqq c$. The discriminant of those is, according to Theorem 5, holomorphic in $|\mu| \leqq c$ and has, not being zero in $\mu = 0$, only a finite number of zeros.

In virtue of Theorem 4, Theorem 5, Theorem 9, and Theorem 12 we note

<u>Theorem 13 :</u> In the neighborhood of an eigenvalue pair (λ_0,μ_0) with $|\mu_0| < \hat{M}$ the equation $\Delta(\lambda,\mu) = 0$ is solved by a finite number of simply or multiply counting power series of $(\mu - \mu_0)^{1/k}$, $k \in \mathbb{N}$, k individual.

If $|\mu_0| < c < \hat{M}$ and if $\mathcal{R}$ is a compact component of $\mathfrak{M}_c$ and $\lambda_0 \in \mathcal{R}$, then each one of the mentioned function elements furnishes by analytic continuation within

$|\mu| < c$ at most a finite number of branch points with a finite value of the function and at most a finite number of branches.

If, for instance,

$$\liminf d_n \cdot \gamma_n^{-1} = \hat{M} = +\infty \quad .$$

Then every solving function element over $\mathbb{C}$ furnishes in the large an algebroid function which is everywhere finite; the exceptional points, that is the μ-projections of all possible branch points, have no finite limiting point. -

We refrain from giving further particulars, for instance, with respect to multiplicities, constant branches, and the appropriate numbering of the eigenvalues λ to μ.

1.1.4. Further presuppositions and conclusions.

For the following considerations we take as a basis further presuppositions in addition to (0) to (6) in 1.1.1..

A supplement to (2) is

(7) For every $\mu \in \mathbb{C}$ and every eigenvalue λ to μ let the order of λ as a zero of $\Delta(\cdot,\mu)$ be equal to the order of the pair $\{F + \mu G + \lambda H, H\}$ and equal to the order of the pair $\{F^* + \mu G^* + \lambda H^*, H^*\}$.

The concept "o r d e r" is here defined as follows: Let $A_o, A_1, A_2, \ldots$ be linear mappings of $\mathcal{L}_1$ into $\mathcal{L}_2$; then for $n = 0,1,2,\ldots$ those subspaces ℓ_n of the elements $c_o \in \mathcal{L}_1$ are considered to which $c_1, c_2, \ldots, c_n \in \mathcal{L}_1$ exist with which there holds

$$\left(A_o + \xi A_1 + \ldots + \xi^n A_n\right)\left(c_o + \xi c_1 + \ldots + \xi^n c_n\right) = \mathcal{O}(\xi^{n+1}) \quad .$$

That is, upon multiplication and collecting terms with the same power of ξ, the coefficients of $\xi^o, \xi^1, \ldots, \xi^n$ are zero. There is obviously

$$\ell_{n+1} \subset \ell_n, \ell_o = \ker A_o \quad .$$

Then we designate

$$\sum_{n=o}^{\infty} \dim \ell_n$$

as the o r d e r of $\{A_o, A_1, A_2, \ldots\}$ and, in particular as the o r d e r of $\{A_o, A_1\}$ if $A_2 = A_3 = \ldots = 0$.

Evidently the order is never smaller than $n(A_o) = \dim \ker A_o$ and $= 0$ if and only if $n(A_o) = 0$. In our case in (7) ℓ_o is just the eigenspace to (λ,μ) of (EP) and of (aEP), respectively.

We consider further a pair $\{A_o, A_1\}$. Then

$$(A_o + \xi A_1)\left(c_o + \xi c_1 + \ldots + \xi^n c_n\right) = \mathcal{O}(\xi^{n+1})$$

is equivalent to

$$A_o c_o = 0,$$
$$A_o c_\nu = -A_1 c_{\nu-1} \qquad (\nu=1,2,\ldots,n)$$

and to

$$(A_o + \xi A_1)\left(c_o + \xi c_1 + \ldots + \xi^n c_n\right) = \xi^{n+1} A_1 c_n .$$

One recognizes easily.

If with $c_o \neq 0$ there is one n with $A_1 c_n = 0$, then the order is ∞ and for all $\xi \in \mathbb{C}$ there holds $\ker (A_o + \xi A_1) \neq 0$.

Conversely:

If the order of $\{A_o, A_1\}$ is finite, then A_1 is injective on the "principal space" consisting of all possible c_ν $(\nu=0,1,2,\ldots)$, whose dimension is equal to the order of $\{A_o, A_1\}$.

Now there follows a complementary presupposition to (6) and (4):

(8) a) Let $|\ |$ be a norm in $\mathcal{R}$ which makes $\mathcal{R}$ a (B)-space.

b) Assume that for $\mu \in \mathbb{C}$, $f \in \mathcal{R}$ and $\Delta(\cdot,\mu) \neq 0$

$$\Delta(\cdot,\mu) HR(\cdot,\mu) f$$

can be continued to an entire analytic function (with values in $(\mathcal{R}, |\ |)$).

c) For the linear mapping $A : H\mathcal{U} \to \mathcal{R}$ there shall be $A \in (1,1;\alpha)$, α and $\alpha\gamma^{(2)}$ relatively bounded, and for $\mu \in \mathbb{C}$, $f \in \mathcal{R}$ and $\Delta(\cdot,\mu) \neq 0$ it is assumed that

$$\Delta(\cdot,\mu) AHR(\cdot,\mu) f$$

can be continued to an entire analytic function (see b)).

d) If $\mathcal{T}$ is the $|\ |$ - closure of $H\mathcal{U}$, then $\|\ \|$ shall be (restricted) norm in $\mathcal{T}$ with

$$\|f\| \leq |f| \qquad (f \in \mathcal{T}) .$$

e) If $\mathcal{T}_A$ is the $|\ |$ - closure of $AH\mathcal{U}$ then $\|\ \|$ shall be (restricted) norm in $\mathcal{T}_A$ with

$$\|f\| \leq |f| \qquad (f \in \mathcal{T}_A) .$$

We notice that (8) d) and e) involve a certain non-degeneracy property for $[\ ,\ ]$: if $f \in \mathcal{T}$ or $f \in \mathcal{T}_A$ and if for all $v \in \mathcal{U}^*$ there is $[f, S^*v] = 0$, then $f = 0$.

We require on the other hand an inverted non-degeneracy condition:

(9) If $v \in \mathcal{U}^*$ and if $[f, S^*v] = 0$ for all $f \in \mathcal{R}$, then there shall be $v = 0$.

Finally we require supplementary to (5):

(10) $\gamma^{(1)}$ and $\gamma^{(2)}$ shall be relatively bounded.

1.1.5. The Residues of the Resolvent. Principal Solutions.

In the following considerations we shall use

Theorem 14: For every $\mu \in \mathbb{C}$ with $|\mu| < \hat{M}$ $\{F + \mu G, H, AH\}$ is "closed" in the following sense: If $u_n \in \mathcal{U}$, f_o, h_o, $a_o \in \mathcal{R}$ and if

$$\|(F + \mu G)u_n - f_o\| \to 0, \quad |Hu_n - h_o| \to 0 ,$$

$$|AHu_n - a_o| \to 0 ,$$

then there exists precisely one $u_o \in \mathcal{U}$ with

$$(F + \mu G)u_o = f_o,\ Hu_o = h_o,\ AHu_o = a_o .$$

Proof: Choose a $\lambda \in \mathbb{C}$ with $m_\gamma(\lambda) > |\mu|$ which is possible according to the definition of $\hat{M}$. If one sets $g_o := f_o + \lambda h_o$ and

$$g_n := (F + \mu G + \lambda H)u_n ,$$

then one has $\|g_n - g_o\| \to 0$ because of (8) d). Now one can write

$$u_n = R(\lambda,\mu)g_n .$$

Then $HR(\lambda,\mu)$ and $AHR(\lambda,\mu)$ are bounded with respect to $\|\ \|$ in the domain and in the range. This follows because of (8) c) and (10) from 1., Theorem 1 and Theorem 3 by choosing there $B := E$, $\beta := 1$ and $A := E$ or A, $\alpha := 1$ or α. Consequently one has with

$$u_o := R(\lambda,\mu)g_o$$

the limiting properties.

$$\|Hu_n - Hu_o\| \to 0 ,\ \|AHu_n - AHu_o\| \to 0 .$$

Now one can again use (8) d) and e). This yields the asserted properties. Uniqueness follwos from (1).

Now we demonstrate

Theorem 15: Let $\mu_o \in \mathbb{C}$, $|\mu_o| < \hat{M}$. Then $\Delta(\cdot,\mu_o) \neq 0$. Let λ_o be an eigenvalue to μ_o and $f \in \mathcal{R}$. Then

$$HR(\lambda,\mu_o)f,\ AHR(\lambda,\mu_o)f$$

which are unique holomorphic $(\mathcal{R},|\ |)$ - valued functions around λ_o, have at λ_o at most a pole and permit with uniquely determined $z_n \in \mathcal{U}$ $(n \geq p \in \mathbb{Z})$ Laurent-expansions around λ_o

$$HR(\lambda,\mu_o)f = \sum_{n=p}^{+\infty} (\lambda - \lambda_o)^n Hz_n ,$$

$$AHR(\lambda,\mu_o)f = \sum_{n=p}^{+\infty} (\lambda - \lambda_o)^n AHz_n ,$$

whereby with $z_{p-1} := 0$

$$(F + \lambda_o H + \mu_o G)z_n = -Hz_{n-1} \qquad (n \neq 0) ,$$

$$(F + \lambda_o H + \mu_o G)z_o = f - Hz_{-1} .$$

Proof: Theorem 3 yields $\Delta(\cdot,\mu_o) \neq 0$. Thus the functions mentioned above have according to (8) b) at most a pole at λ_o. Therefore one can with $x_n \in \mathcal{R}$, $w_n \in \mathcal{R}$ expand around λ_o

$$HR(\lambda,\mu_o)f = \sum_{n=p}^{+\infty} (\lambda - \lambda_o)^n x_n ,$$

$$AHR(\lambda,\mu_o)f = \sum_{n=p}^{+\infty} (\lambda - \lambda_o)^n w_n .$$

Moreover, one takes notice of

$$(F + \lambda_o H + \mu_o G)R(\lambda,\mu_o)f - f = -(\lambda - \lambda_o)HR(\lambda,\mu_o)f = -\sum_{n=p}^{+\infty} (\lambda - \lambda_o)^n x_{n-1}$$

where $x_{p-1} := 0$.

In all three cases the coefficients are expressed by integrals along an appropriate circle $\mathcal{R}$ around λ_o :

$$\frac{1}{2\pi i}\int_{\mathcal{R}} (\lambda-\lambda_o)^{-n-1}\, HR(\lambda,\mu_o)f d\lambda = x_n \quad ,$$

$$\frac{1}{2\pi i}\int_{\mathcal{R}} (\lambda-\lambda_o)^{-n-1}\, AHR(\lambda,\mu_o)f d\lambda = w_n \quad ,$$

$$\frac{1}{2\pi i}\int_{\mathcal{R}} (\lambda-\lambda_o)^{-n-1}[(F+\lambda_o H+\mu_o G)R(\lambda,\mu_o)f - f]d\lambda = -x_{n-1} \quad .$$

Here one passes over jointly to refining approximating sums and obtains sequences $v_{nm} \in \mathcal{U}$ and $\alpha_{nm} \in \mathbb{C}$ with

$$|Hv_{nm} - x_n| \to 0 \ , \ |AHv_{nm} - w_n| \to 0 \ ,$$

$$\alpha_{nm} \to \delta_{no} = \begin{cases} 0 & (n \neq 0) \\ 1 & (n = 0) \end{cases} \ ,$$

$$|[(F+\lambda_o H+\mu_o G)v_{nm} - \alpha_{nm}f] + x_{n-1}| \to 0$$

for $m \to \infty$. Since all integrands are in $H\mathcal{U}$ or $AH\mathcal{U}$, respectively, so do also the terms of the approximating sums. Due to (8) d) and e) the convergences also hold in all cases with $\|\ \|$ instead of $|\ |$. Now one chooses $\lambda' \in \mathbb{C}$ with $\Delta(\lambda',\mu_o) \neq 0$ and solves - see (6) -

$$(F+\lambda' H+\mu_o G)v_o = f,\ v_o \in \mathcal{U} \ .$$

Then one can introduce

$$u_{nm} := v_{nm} - \alpha_{nm}v_o \in \mathcal{U}$$

and obtains

$$\|(F+\lambda' H+\mu_o G)u_{nm} - (-x_{n-1} + (\lambda'-\lambda_o)x_n)\| \to 0 \ ,$$

$$|Hu_{nm} - (x_n - \delta_{no}Hv_o)| \to 0 \ ,$$

$$|AHu_{nm} - (w_n - \delta_{no}AHv_o)| \to 0 \ .$$

Here Theorem 14 can be applied. It says that there exists a unique $u_n \in \mathcal{U}$ with

$$Hu_n = x_n - \delta_{no}\, Hv_o \ ,$$

$$AHu_n = w_n - \delta_{no}\, AHv_o \ ,$$

$$(F+\lambda' H+\mu_o G)u_n = -x_{n-1} + (\lambda'-\lambda_o)x_n \quad .$$

With $z_n := u_n + \delta_{no}v_o \in \mathcal{U}$ the theorem is proved. -

In the following we introduce the practical abbreviations

$$\langle u,v\rangle := [Hu,S^*v] \qquad (u \in \mathcal{U}\ ,\ v \in \mathcal{U}^*).$$

Moreover, we write for the z_{-1} in Theorem 15

$$z_{-1} =: r(\lambda_o,\mu_o)f$$

and clarify in the following the structure of the linear mapping

$$r(\lambda_o,\mu_o) : \mathcal{R} \to \mathcal{U}$$

which is thus induced.

Theorem 16 : Let $\mu_o \in \mathbb{C}$ with $|\mu_o| < \hat{M}$ and let the eigenvalue λ_o to μ_o be a k-fold zero of $\Delta(\cdot,\mu_o)$.

a) Then after (7) there exists a basis

$$h_1, h_2, \ldots, h_k$$

of the principal space $\mathfrak{H}$ of $\{F + \lambda_o H + \mu_o G, H\}$ and a basis

$$h_1^*, h_2^*, \ldots, h_k^*$$

of the principal space $\mathfrak{H}^*$ of $\{F^* + \lambda_o H^* + \mu_o G^*, H^*\}$ which form a normalized biorthogonal system in the sense

$$\langle h_\nu, h_\mu^* \rangle = \delta_{\nu\mu} \qquad ((\nu,\mu) \in \{1,2,\ldots,k\}^2) \ .$$

b) With respect to each such biorthogonal basis there holds the representation

$$r(\lambda_o,\mu_o)f = \sum_{\kappa=1}^{k} [f,S^*h_\kappa^*]h_\kappa \qquad (f \in \mathcal{R}) \ .$$

c) There exists a biorthogonal basis for $\mathfrak{H}$, $\mathfrak{H}^*$ such that with a division in groups by suitable numbers

$$0 = k_o < k_1 < \ldots < k_{r-1} < k_r = k$$

there holds

$$(F + \lambda_o H + \mu_o G)h_\nu = \begin{cases} 0 & (\nu=k_\rho+1, \rho=0,1,\ldots,r-1) \\ -Hh_{\nu-1} & (\text{else}) \ , \end{cases}$$

$$(F^* + \lambda_o H^* + \mu_o G^*)h_\nu^* = \begin{cases} 0 & (\nu=k_\rho, \rho=1,2,\ldots,r) \\ -H^*h_{\nu+1}^* & (\text{else}) \ . \end{cases}$$

Proof: The comparison of coefficients in Theorem 15 yields

(*) $$z_{-1} = r(\lambda_o,\mu_o)f \in \mathfrak{H} \ .$$

On the other hand there is

$$(F + \lambda_o H + \mu_o G)z_o = f - Hz_{-1} \ ,$$

$$(F + \lambda_o H + \mu_o G)z_n = - Hz_{n-1} \qquad (n \in \mathbb{N}) \ .$$

With that one verifies for an $h^* \in \mathfrak{H}^*$ with

$$(F^* + \lambda_o H^* + \mu_o G^*)h^* = -H^*u_1^*$$

$$(F^* + \lambda_o H^* + \mu_o G^*)u_\nu^* = -H^*u_{\nu+1}^* \qquad (\nu=1,2,\ldots,k)$$

$$u_{k+1}^* = 0$$

that

$$[f - Hz_{-1}, S^*h^*] = [Sz_o, (F^* + \lambda_o H^* + \mu_o G^*)h^*] =$$

$$= [Sz_o, -H^* u_1^*] = [-Hz_o, S^* u_1^*] =$$

$$= [(F + \lambda_o H + \mu_o G)z_1, S^* u_1^*] = \dots =$$

$$= [Sz_k, -H^* u_{k+1}^*] = 0 .$$

Therefore

$$(\times) \qquad [f, S^* h^*] = \langle z_{-1}, h^* \rangle \qquad (h^* \in \mathfrak{H}^*) .$$

Now (9) is applied. Then $(\times)$ with $(*)$ yields: to every $0 \neq h^* \in \mathfrak{H}^*$ there exists an $h \in \mathfrak{H}$ with

$$\langle h, h^* \rangle \neq 0 .$$

Because of $\dim \mathfrak{H} = \dim \mathfrak{H}^* = k$ - see (7) - there holds conversely: to every $0 \neq h \in \mathfrak{H}$ there exists an $h^* \in \mathfrak{H}^*$ with

$$\langle h, h^* \rangle \neq 0 .$$

For otherwise one could complete h by $h_2, h_3, \dots, h_k$ to a basis of $\mathfrak{H}$ and find $0 \neq h^* \in \mathfrak{H}^*$ with

$$\langle h_\kappa, h^* \rangle = 0 \qquad (\kappa = 2, 3, \dots, k) ,$$

because these are $k-1$ equations for $h^* \in \mathfrak{H}^*$ with $\dim \mathfrak{H}^* = k$. But then one would have $\langle u, h^* \rangle = 0$ for all $u \in \mathfrak{H}$ which is in contradiction to what we have shown above. Therefore $\langle\ ,\ \rangle$ on $\mathfrak{H} \times \mathfrak{H}^*$ is (in every direction) not degenerate, one can find biorthogonal bases according to a), and $(\times)$ proves immediately statement b). Thus $r(\lambda_o, \mu_o)$ turns out to be a linear idempotent surjective mapping. Statement c) follows from considering $A := H^{-1}(F + \lambda_o H + \mu_o G)$ on $\mathfrak{H}$ and $A^* := H^{*-1}(F^* + \lambda_o H^* + \mu_o G^*)$ on $\mathfrak{H}^*$, which are mutually adjoint with respect to $\langle\ ,\ \rangle$. Then one determines a basis $h_1, h_2, \dots, h_k$ with the mentioned properties which corresponds to the Jordan normal form of the matrix representation of A. If one constructs the dual basis $h_1^*, h_2^*, \dots, h_k^*$ of $\mathfrak{H}^*$ with respect to $\langle\ ,\ \rangle$, then A^* in this basis must correspond to the transpose of the Jordan normal form for A. These are the properties put down for $h_1^*, \dots, h_k^*$.

In particular there results

<u>Theorem 17 :</u> To $(\lambda_n, 0)$ there exist eigensolutions y_n of (EP) with

$$\langle y_n, y_m^* \rangle = \delta_{nm} \qquad ((n,m) \in \mathbb{N}^2) .$$

By the way, a simpler direct proof can be given for this Theorem.

For the sake of completeness we add

<u>Theorem 18 :</u> Let $|\mu_o| < \hat{M}$ and let h, h^* be principal vector and adjoint principal vector to different eigenvalues λ and λ', respectively. Then

$$\langle h, h^* \rangle = 0 .$$

<u>Proof :</u> With $h_o := h, h_o^* := h^*$ let

$$(F + \lambda H + \mu_o G) h_\nu = H h_{\nu+1} \qquad (\nu = 0, 1, \dots, n-1),\ h_n = 0 ,$$

$$(F^* + \lambda' H^* + \mu_o G^*) h_\kappa^* = H^* h_{\kappa+1}^* \qquad (\kappa = 0, 1, \dots, m-1),\ h_m^* = 0 .$$

Then one works out

$$\langle h_{\nu+1}, h^*_\kappa \rangle = [(F + \lambda H + \mu_0 G)h_\nu, S^*h^*_\kappa] =$$
$$= [Sh_\nu, (F^* + \mu_0 G^* + \lambda H^*)h^*_\kappa] =$$
$$= [Sh_\nu, H^*h^*_{\kappa+1} + (\lambda - \lambda')H^*h^*_\kappa]$$

which is equivalent to

$$(\lambda - \lambda') \langle h_\nu, h^*_\kappa \rangle = \langle h_{\nu+1}, h^*_\kappa \rangle - \langle h_\nu, h^*_{\kappa+1} \rangle \ .$$

By iteration (or by induction with respect to the sum of the orders) the assertion is obtained.

Theorem 19 : Let $|\mu_0| < \hat{M}$ and let $h_\nu \neq 0$ $(\nu=1,2,\ldots,n)$ be principal vectors to different $\lambda_1(\mu_0),\ldots,\lambda_n(\mu_0)$. Then the Hh_ν , consequently also the h_ν , are linearly independent.

Proof : It suffices to assume

$$Hh_1 + Hh_2 + \ldots + Hh_n = 0 \ .$$

Then one chooses after Theorem 16 an adjoint principal vector h_1^* to $(\lambda_1(\mu_0),\mu_0)$ with $\langle h_1, h_1^* \rangle \neq 0$ and obtains with Theorem 16 a)

$$0 = \sum_{\nu=1}^{n} [Hh_\nu, Sh_1^*] = \sum_{\nu=1}^{n} \langle h_\nu, h_1^* \rangle = \langle h_1, h_1^* \rangle$$

This is a contradiction .

In the following we assume:

Let $(\mathcal{R}_1, \sqcap)$ be a pseudonormalized space and an extension of $(\mathcal{R}, \sqcap)$; let $\mathcal{R}$ be dense in $\mathcal{R}_1$ with respect to $\sqcap$.

(If $\sqcap$ is norm in $\mathcal{R}$, then it is known that there exists a (B)-space closure $(\mathcal{R}_1, \sqcap)$ which is unique apart from an isomorphism.)

We show

Theorem 20 : Under the presuppositions

(i) $0 \leq |\mu_0| \leq c < \hat{M}$, λ_0 is eigenvalue to μ_0 ;

(ii) $\mathfrak{M}_c$ decomposes into a countable number of compact components;

(iii) $B \in (II,2;\beta)$, $\beta\delta$ and $\beta\gamma^{(1)}\delta$ are subdominant;

(iv) h^* is principal vector to $\{F^* + \mu_0 G^* + \lambda_0 H^*, H^*\}$;

$[Bf, S^*h^*]$ $(f \in \vartheta_B)$ depends $\sqcap$-continuous on f. Therefore $[B\cdot, S^*h^*]$ can be extended uniquely linear and $\sqcap$-continuous to the $\sqcap$-closure $\overline{\vartheta_B}^1$ of ϑ_B in $\mathcal{R}_1$.

Proof : Let $\lambda_o \in \mathfrak{K}$, $\mathfrak{K}$= compact component of $\mathfrak{M}_c$, and let $\mathcal{L}$ be a closed rectifiable contour which around $\mathfrak{K}$ and around $\mathfrak{M}_c \setminus \mathfrak{K}$ has the circuital numbers 1 and 0 , respectively, and on which $m_\gamma(\lambda) \geq c' > c$. Now one considers

$$(\times) \qquad \frac{1}{2\pi i} \int_{\mathcal{L}} HR(\lambda,\mu_0)Bf d\lambda = \Sigma Hr(\lambda',\mu_0)Bf \quad ,$$

with the finite sum extending over the different zeros λ' of $\Delta(\cdot,\mu_0)$ in $\mathfrak{K}$. All pictures of $f \in \vartheta_B$ are in the space

$$\Sigma\, Hr(\lambda', \mu_o)\mathcal{R} \quad ,$$

which has according to Theorem 16 and Theorem 19 as its dimension just the sum of the orders of the zeros. In the left member of (×) the norm $\| \ \|$ of the contour integral can, according to Theorem 1 and Theorem 3, be estimated by $C \cdot \lceil f \rceil$ with a constant C which is independent of f. Thereby

$$\Sigma\, Hr(\lambda', \mu_o) B$$

is recognized as a $\lceil\,\rceil$ - continuous linear mapping of ϑ_B into the mentioned finite-dimensional space. But this implies that in any basis representation according to Theorem 16, a),b) the coefficients are $\lceil\,\rceil$ - continuous. The conclusion on the extensibility is well known.

1.1.6. Equiconvergence.

In the previous sections all facts for the formulation of the following main theorem have been made available.

Theorem 21 : Let the following presuppositions hold:

(i) $0 \leq |\mu_o| < c < \hat{M}$;

(ii) $B \in (II,2;\beta)$, $\beta\delta$ and $\beta\gamma^{(1)}\delta$ are subdominant;

(iii) A,α satisfy (8) c); moreover, $A \in (I,2;\alpha)$ and $\alpha\gamma^{(2)}\eta$ is subdominant;

(iv) there exist finite sets $\mathbb{N}_k \neq \emptyset$ with

$$\mathbb{N}_1 \subset \mathbb{N}_2 \subset \mathbb{N}_3 \subset \ldots \uparrow \mathbb{N}$$

and chains K_k of closed, continuous and rectifiable contours with

$$m_\gamma(\lambda) \geq c \qquad (\lambda \in (K_k)) \ ,$$

$$u(K_k, \lambda_n) = \begin{cases} 1 & (n \in \mathbb{N}_k) \\ 0 & (n \in \mathbb{N} \setminus \mathbb{N}_k) \ , \end{cases}$$

$$\int_{K_k} \sigma_{\alpha\gamma^{(2)}\eta}(\lambda)\ \sigma_{\beta\gamma^{(1)}\delta}(\lambda)\,|d\lambda| \to 0 \qquad (k \to \infty) \ .$$

Then the eigenvalues $\lambda_n(\mu_o)$ $(n \in \mathbb{N})$ to μ_o, counted according to their order, can be so numbered that

$$u(K_k, \lambda_n(\mu_o)) = \begin{cases} 1 & (n \in \mathbb{N}_k) \\ 0 & (n \in \mathbb{N} \setminus \mathbb{N}_k) \end{cases} \ .$$

To these eigenvalues there exist principal vectors $y_n(\mu_o)$ and adjoint principal vectors $y_n^*(\mu_o)$ with

$$\langle y_n(\mu_o),\ y_m^*(\mu_o)\rangle = \delta_{nm} \qquad ((n,m) \in \mathbb{N}^2)$$

such that the principal vectors belonging to one and the same eigenvalue form bases according to Theorem 16, a),b),c).

If one chooses $C < \infty$, then there holds uniformly for $f \in \overline{\vartheta}_B^1$ (see Theorem 20) with $\overline{|f|} \leq C$

$$\left| \sum_{n \in \mathbb{N}_k} \left(\left[Bf, S^* y_n^*(\mu_o) \right] AHy_n(\mu_o) - \left[Bf, S^* y_n^* \right] AHy_n \right) \right|_o \longrightarrow 0 \qquad (k \to \infty) .$$

Proof: The possibility to number the $\lambda_n(\mu_o)$ and to choose the $y_n(\mu_o), y_n(\mu_o)^*$ in this way is evident due to 1.1.3. and Theorem 16 and Theorem 18. The y_n, y_n^* are chosen with Theorem 17. The sum, which is now to be estimated, is represented for $f \in \vartheta_B$ with the residue theorem in accordance with (8) c), Theorem 15, Theorem 16 by

$$\frac{1}{2\pi i} \int_{K_k} AH(R(\lambda,\mu_o) - R(\lambda,0))Bfd\lambda .$$

This is estimated with Theorem 3, (3.4) by

$$\frac{|\mu_o|}{2\pi} \quad \frac{\overline{|f|}}{1-|\mu_o|c^{-1}} \int_{K_k} \sigma_{\alpha\gamma^{(2)}\eta}(\lambda)\ \sigma_{\beta\gamma^{(1)}\delta}(\lambda)\,|d\lambda| \quad .$$

With Theorem 20 this estimate is extended to $\overline{\vartheta}_B^1$ and the assertion is verified.

We remark that Theorem 21 is wholly tailored to Theorem 3, (3.4). Obviously the remaining estimates of Theorem 3 can be used in an analogous way. But we resign here an explicit formulation of the resulting theorems.

1.1.7. Holomorphy Properties. Estimates.

Here we assume in additon

(11) $$| \ |_o = | \ |$$

and use the following notations and considerations:

Let ϑ be a subspace of $\mathcal{R}$,

$$\|f\| < \infty \qquad (f \in \vartheta) ,$$

and $\mathcal{L}(\vartheta)$ the space of linear mappings of ϑ into $\mathcal{R}$. For $L \in \mathcal{L}(\vartheta)$ we define the pseudonorms

$$\|L\| := \min \{c \geq 0 : \|Lf\| \leq c\,\|f\|,\ f \in \vartheta\} ,$$

$$|L| := \min \{c \geq 0 : |Lf| \leq c\,\|f\|,\ f \in \vartheta\}$$

with $\min \emptyset := +\infty$. With

$$\mathcal{L}_b(\vartheta) := \{L \in \mathcal{L}(\vartheta) : |L| < \infty\}$$

$(\mathcal{L}_b(\vartheta), |\ |)$ becomes a (B) - space.

At first we show - by the way without using (7) , (8) , (9) -

Theorem 22 : Let

(i) $B \in (II,1;\beta);\beta$ and $\beta\gamma^{(1)}$ relatively bounded;

$\|f\| < \infty \quad (f \in \vartheta_B)$;

(ii) $A \in (I,2;\alpha)$; $\alpha\eta, \alpha\gamma^{(2)}r, \alpha\beta r$ subdominant.

Then

$$AHR(\lambda,\mu)B$$

defines a holomorphic function of two variables in the open set $\{(\lambda,\mu) \in \mathbb{C}^2 : m_\gamma(\lambda) > |\mu|\}$ with values in the (B)-space $(\mathcal{L}_b(\vartheta_B),|\ |)$.

Proof : The two-parametric resolvent equation

$$R(\lambda,\mu) = R(\lambda_o,\mu_o) - (\lambda-\lambda_o)R(\lambda_o,\mu_o)HR(\lambda,\mu) - (\mu-\mu_o)R(\lambda_o,\mu_o)GR(\lambda,\mu)$$

is iterated and combined with AH and B :

$$\begin{aligned} AHR(\lambda,\mu)B = {} & AHR(\lambda_o,\mu_o)B - (\lambda-\lambda_o)AHR(\lambda_o,\mu_o)HR(\lambda_o,\mu_o)B - \\ & - (\mu-\mu_o)AHR(\lambda_o,\mu_o)GR(\lambda_o,\mu_o)B + \\ & + (\lambda-\lambda_o)^2 AHR(\lambda_o,\mu_o)HR(\lambda_o,\mu_o)HR(\lambda,\mu)B + \\ & + (\mu-\mu_o)^2 AHR(\lambda_o,\mu_o)GR(\lambda_o,\mu_o)GR(\lambda,\mu)B + \\ & + (\lambda-\lambda_o)(\mu-\mu_o)AHR(\lambda_o,\mu_o)HR(\lambda_o,\mu_o)GR(\lambda,\mu)B + \\ & + (\lambda-\lambda_o)(\mu-\mu_o)AHR(\lambda_o,\mu_o)GR(\lambda_o,\mu_o)HR(\lambda,\mu)B \ . \end{aligned}$$

Then one applies Theorem 1, Theorem 3, and Theorem 3a and uses repeatedly the thought that

$$|L_1 \circ L_2| \leq |L_1| \cdot \|L_2\| \quad ,$$

$$\|L_1 \circ L_2\| \leq \|L_1\| \cdot \|L_2\| \quad ,$$

provided everything is defined. Thus one recognizes, with the made presuppositions including (10) , that all terms within the mentioned set are finite with respect to $|\ |$, or are uniformly $|\ |$ - bounded for (λ,μ) in a neighborhood of a fixed (λ_o,μ_o). From this result one can read off the assertion and even the partial derivatives. -

With the help of (8) the statement of Theorem 22 can be extended.

Theorem 23 : Let B and the A given in (8) satisfy the presuppositions (i), (ii) of Theorem 22 . Let $0 < c < \hat{M}$ and let $\mathcal{R}$ be a compact component of $\mathfrak{M}_c$.

Then

$$\Delta(\lambda,\mu)AHR(\lambda,\mu)B$$

is defined for $m_\gamma(\lambda) > |\mu|$ as well as in a neighborhood of $\mathcal{R} \times \{|\mu| \leq c\}$ and is there a holomorphic function of the two variables λ,μ with values in the (B)-space $(\mathcal{L}_b(\vartheta_B),|\ |)$.

Proof : There exists a closed, continuous, and rectifiable contour, which runs once around $\mathcal{R}$ in positive sense and does not run around $\mathfrak{M}_c \setminus \mathcal{R}$, and on which there is $m_\gamma(\lambda) \geq c' > c$. For $f \in \vartheta_B$, λ near $\mathcal{R}$ "within" $\mathcal{L}$ and $|\mu| < c'$ there follows from (8) and with the help of Cauchy's integral theorem the representation

$$\Delta(\lambda,\mu)AHR(\lambda,\mu)Bf = \frac{1}{2\pi i} \int_{\mathcal{L}} \frac{\Delta(\lambda',\mu)}{\lambda' - \lambda} AHR(\lambda',\mu)Bfd\lambda' \quad .$$

Now in the right member also the operator integral exists with respect to $|\ |$ according to Theorem 22. Therefore a well-known argument permits to cancel f in both members. Thus the additional statement of Theorem 23 can be read off in an elementary way. - In addition to Theorem 23 we remark that, for instance with the principle of the maximum, one can here obtain estimates also for $R(\lambda,\mu)$ analogous to Theorem 1 and Theorem 3 without assuming $m_\gamma(\lambda) > |\mu|$. It would even be possible to replace $\Delta(\cdot,\mu)$ by suitable polynomials. But this idea will here not be further pursued.

Subsequently to Theorem 23 one can - using all presuppositions - consider the residues of 1.1.5..

Theorem 24 : Let $0 < c < \hat{M}$ and let $\mathcal{R}$ be a compact component of $\mathfrak{M}_c$. B and the A given in (8) shall satisfy the presuppositions (i), (ii) of Theorem 22. Then

$$\sum_{\lambda'} AHr(\lambda',\mu)B ,$$

ranging over the finite number of values $\lambda' \in \mathcal{R}$ with $\Delta(\lambda',\mu) = 0$, furnishes a holomorphic function in $|\mu| \leqq c$ with values in the (B)-space $(\mathcal{L}_b(\mathcal{V}_B),|\ |)$.

Proof : One considers again a closed continuous rectifiable contour $\mathcal{L}$ which runs once around $\mathcal{R}$ in positive sense and does not run around $\mathfrak{M}_c \setminus \mathcal{R}$ and on which $m_\gamma(\lambda) \geqq c' > c$. Then with 1.1.5. and Theorem 23 the above sum is written in the form

$$\frac{1}{2\pi i}\int_{\mathcal{L}} AHR(\lambda,\mu)Bd\lambda .$$

From this one can, using Theorem 22, read off the assertion in a known manner. -

It remains to specialize to the case treated in 1.1.3., Theorem 7, Theorem 10 using the same notations.

Theorem 25 : B and the A,α given in (8) shall satisfy the presuppositions (i), (ii) of Theorem 22. Let $\lambda_{m_1} = \lambda_{m_2} = \ldots = \lambda_{m_k}$ be a k-fold zero of $\Delta(\cdot,0)$. Let $|\mu| < r_{m_1}$ $(= r_{m_2} = \ldots = r_{m_k})$. Let $y_{m_\kappa}(\mu)$, $y^*_{m_\kappa}(\mu)$ $(\kappa=1,2,\ldots,k)$ be with respect to $\langle\cdot,\cdot\rangle$ a biorthogonal and normalized system of principal solutions and adjoint principal solutions to the eigenvalues $\lambda_{m_\kappa}(\mu)$ which are counted according to their order with $|\lambda_{m_\kappa}(\mu) - \lambda_{m_\kappa}| \leqq \gamma_{m_\kappa}|\mu|$. Then

$$\sum_{\kappa=1}^{k}\left[B\cdot\ ,\ S^*y^*_{m_\kappa}(\mu)\right]AHy_{m_\kappa}(\mu) = \sum_{\ell=0}^{\infty}\mu^\ell P_{m_1,\ell}$$

furnishes a function which is holomorphic for $|\mu| < r_{m_1}$ and has values in the (B)-space $(\mathcal{L}_b(\mathcal{V}_B),|\ |)$.

The special case $k = 1$, $m_1 := m$ and $B := id_{H\mathcal{U}}$, with which (i) in Theorem 22 is obviously satisfied, is of particular interest. If then A is suitable according to (ii), the above mappings on Hy_m with the eigensolution y_m to $(\lambda_m,0)$ can be applied. One obtains in $(\mathcal{R},|\ |)$ for $|\mu| < r_m$

(*) $$\langle y_m, y_m^*(\mu)\rangle\, AHy_m(\mu) = \sum_{\ell=0}^{\infty} \mu^{\ell} c_{m\ell}$$

with

(**) $$c_{mo} = AHy_m,\ c_{m\ell} = P_{m\ell}Hy_m \ .$$

1.1.8. Additional estimates.

In the following we give some supplementary estimates, on the one hand to Theorem 21 , on the other hand to Theorem 25 and (*) . We lean always on (0) to (10) .

Theorem 26 : Let A,α satisfy (8) c). In addition let $A \in (I,2;\alpha)$ and $\alpha\gamma^{(2)}\eta$ be subdominant; further assume $B \in (II,1;\beta)$, $\beta\gamma^{(1)}$ relatively bounded, and $f \in \vartheta_B$. For $|\mu| < r_{m_1}$

$$\lambda_{m_\kappa}, \lambda_{m_\kappa}(\mu),\ y_{m_\kappa}(\mu),\ y^*_{m_\kappa}(\mu) \qquad (\kappa=1,2,\ldots,k)$$

shall be given and designated as in Theorem 25.

If

$$\gamma_{m_1}\left(= \gamma_{m_2} = \ldots = \gamma_{m_k}\right) > 0$$

and

$$C_{m_1} := \gamma_{m_1} r_{m_1} \cdot \max\left\{\sigma_{\alpha\gamma^{(2)}\eta}(\lambda) M_{\beta\gamma^{(1)}}(\lambda) : |\lambda-\lambda_{m_1}| \leqq \gamma_{m_1} r_{m_1}\right\} ,$$

then

$$\left|\sum_{\kappa=1}^{k}\left(\left[Bf, S^* y^*_{m_\kappa}(\mu)\right] AHy_{m_\kappa}(\mu) - \left[Bf, S^* y^*_{m_\kappa}\right] AHy_{m_\kappa}\right)\right|_0 \leqq$$

$$\leqq C_{m_1} \frac{|\mu|}{1-|\mu| r_{m_1}^{-1}} \|f\| .$$

Proof : The considered difference can be written

$$\frac{1}{2\pi i}\int_{|\lambda-\lambda_{m_1}| = \gamma_{m_1} r_{m_1}} AH(R(\lambda,\mu) - R(\lambda,0))Bf d\lambda$$

and estimated with Theorem 3, (3.3). As a consequence - now with (11) - we obtain:

Theorem 27 : In Theorem 25 one has for $\ell \in \mathbb{N}$ with $0^0 := 1$

$$|P_{m_1,\ell}| \leqq C_{m_1} \ell \left(\frac{\ell-1}{\ell} r_{m_1}\right)^{-(\ell-1)}$$

and correspondingly in (*)

$$|c_{m\ell}| \leqq C_m \|Hy_m\| \, \ell\left(\frac{\ell-1}{\ell} r_m\right)^{-(\ell-1)} .$$

This (rough) estimate results from an application of the Cauchy coefficients formule with optimal radius.

We remark that obviously

$$\|Hy_m\| = \|(\delta_{mn})_{n\in\mathbb{N}}\|_2 ,$$

that is $=1$ in many applications.

From Theorem 27 one reads off

Theorem 28 : If all λ_m apart from a finite number are simple zeros of $\Delta(\cdot,0)$, if $r_m \to \infty$, and if, for instance, C_m or $C_m \|Hy_m\|$, respectively, are bounded, then the power series in Theorem 25 and (*) are at the same time asymtotic series for $m \to \infty$.

The estimates with $\lceil f \rceil$ in place of $\|f\|$ are of a similar kind.

Theorem 29 : Let A,α satisfy (8) c); in addition let $A \in (I,2;\alpha)$ and $\alpha\gamma^{(2)}\eta$ be subdominant. Let $B \in (II,2;\beta),\beta\delta$ and $\beta\gamma^{(1)}\delta$ be subdominant.

$$\lambda_{m_\kappa}, \lambda_{m_\kappa}(\mu),\ y_{m_\kappa}(\mu),\ y^*_{m_\kappa}(\mu) \qquad (\kappa=1,2,\ldots,k)$$

shall be given and designated as in Theorem 25. Let $|\mu| < r_{m_1}$. Then for $f \in \overline{\vartheta_B}^{-1}$ (see Theorem 20) with

$$D_{m_1} := \gamma_{m_1} r_{m_1} \cdot \max\left\{\sigma_{\alpha\gamma^{(2)}\eta}(\lambda)\sigma_{\beta\gamma^{(1)}\delta}(\lambda) \ :\ |\lambda-\lambda_{m_1}| = \gamma_{m_1} r_{m_1}\right\}$$

there holds the estimate

$$\left|\sum_{\kappa=1}^{k}\left(\left[Bf,S^*y^*_{m_\kappa}(\mu)\right] AHy_{m_\kappa}(\mu) - \left[Bf,S^*y^*_{m_\kappa}\right] AHy_{m_\kappa}\right)\right|_0 \leq$$

$$\leq D_{m_1} \frac{|\mu|}{1-|\mu| r^{-1}_{m_1}} \lceil f \rceil .$$

The proof is analogous to that for Theorem 26 with Theorem 3, (3.4).

We refrain here again from giving further estimates and conclusions.

1.1.9. On the application to boundary value problems for ordinary differential equations and differential systems.

In particular, very general "adjoint" boundary eigenvalue problems for ordinary differential equations and differential systems, respectively, can be treated in the form 1.1.1., (0) . Here one can show that with weak assumptions there exists always a Δ with (2) and (7) .

For that purpose we send on in advance a theorem on matrices resuming the general definitions of the ℓ_n and of the order in connection with 1.1.4. , (7) .

Theorem 30 : Let

$$D(\lambda) = D_0 + \lambda D_1 + \lambda^2 D_2 + \ldots$$

be a complex (n,n)-matrix which is holomorphic around $\lambda = 0$. Then the order of $\{D_0,D_1,D_2,\ldots\}$ is equal to the order of 0 as zero of

$$\Delta(\lambda) := \det D(\lambda) .$$

Proof : Only for $\Delta(0) = 0$ or $\ker D_o \neq \{0\}$ there is something to be proved. One constructs an (n,n) - matrix C_{oo} whose first $r_o := \dim \mathcal{E}_o$ columns form a basis of $\mathcal{E}_o$, while the last $n - r_o$ columns are zero, and in addition a matrix C_{o1}, whose first r_o columns are zero in such a manner that $C_{oo} + C_{o1}$ is invertible. Then obviously $D(\lambda)\,(C_{oo} + C_{o1})$ is an (n,n) - matrix, which is holomorphic around $\lambda = 0$, and whose first r_o columns vanish for $\lambda = 0$. Now one extracts a factor λ from $D(\lambda)C_{oo}$ and defines a new matrix

$$D_1(\lambda) = \frac{1}{\lambda} D(\lambda)\ (C_{oo} + \lambda C_{o1}) .$$

Then the order of 0 as a zero of $\det D_1(\lambda)$ is just by r_o smaller than of $\det D(\lambda)$. We introduce now the

Lemma : If $p \in \mathbb{N}$ and if with $(n,1)$ vectors c_ν

$$(*) \qquad D(\lambda)(c_o + \lambda c_1 + \ldots + \lambda^p c_p) = \mathcal{O}(\lambda^{p+1}) ,$$

then there exist unique $d_o, \ldots, d_{p-1}, C_{oo}d_p$ such that

$$(**) \qquad c_o + \lambda c_1 + \ldots + \lambda^p c_p = (C_{oo} + \lambda C_{o1})\left(d_o + \lambda d_1 + \ldots + \lambda^{p-1} d_{p-1}\right) + \lambda^p C_{oo} d_p$$

and one has

$$(*\!*\!*) \qquad D_1(\lambda)\left(d_o + \lambda d_1 + \ldots + \lambda^{p-1} d_{p-1}\right) = \mathcal{O}(\lambda^p) .$$

Conversely, if $(*\!*\!*)$ holds with $(n,1)$ - vectors d_ν and if the c_ν $(\nu=0,1,\ldots,p)$ are defined by $(**)$ with d_p arbitrary, then $(*)$ holds. $d_o = 0$ holds if and only if $c_o = 0$.

Proof : $(**)$ is equivalent to

$$\begin{aligned} c_o &= C_{oo} d_o \\ c_1 &= C_{oo} d_1 + C_{o1} d_o \\ &\vdots \\ c_p &= C_{oo} d_p + C_{o1} d_{p-1} . \end{aligned}$$

Since $D_o c_o = 0$ the first equation determines uniquely the first r_o elements of d_o ; the second equation determines uniquely the last $n - r_o$ elements of d_o and the first r_o elements of d_1 etc. . From the definition of D_1 and from $D_o C_{oo} = 0$ follows $(*\!*\!*)$. The inversion is trivial, due to the definition of C_{oo}. Of course, $d_o = 0$ yields $c_o = 0$. If, conversely, $c_o = 0$ then $D_o c_1 = 0$ according to $(*)$. According to the definition of C_{oo} the second equation gives $C_{o1} d_o = 0$ and together with the first one $d_o = 0$.

Because of linearity the last consideration implies

$$\dim \mathcal{E}_p = \dim \mathcal{E}^1_{p-1}$$

with the $\mathcal{E}^1_\nu$ being the corresponding spaces for $D_1(\lambda)$. This and the above remark about the order of the zeros of the determinants yields inductively the proof of Theorem 30.

We consider now boundary-eigenvalue problems for systems of ordinary differential equations of the first order (which permits the incorporation of differential equations of higher order), and that in the real and in the complex domain at the same time.

For that let F_{o1}, H_{o1}, F_{o2}, H_{o2} be continuous functions in a non-trivial interval $i \subset \mathbb{R}$ or holomorphic functions in a simply connected domain $\mathcal{G} \in \mathbb{C}$, respectively, with values in the (n,n)-matrices with complex elements. F_{11}, H_{11}, F_{12}, H_{12} shall be (constant) complex (n,n)-matrices. a and b are different points in i or $\mathcal{G}$, respectively. With $\mathcal{U}_o$ being the space $C_1(i,\mathbb{C}^n)$ of continuously differentiable functions with values in $\mathbb{C}^n$, or the space $H(\mathcal{G},\mathbb{C}^n)$ of functions which are holomorphic in $\mathcal{G}$ and have values in $\mathbb{C}^n$, respectively, we define with $y \in \mathcal{U}_o$ linear mappings

$$(F_o y)(x) := F_{o1}(x)y'(x) + F_{o2}(x)y(x) ,$$

$$(H_o y)(x) := H_{o1}(x)y'(x) + H_{o2}(x)y(x)$$

in $C_o(i,\mathbb{C}^n)$ or in $H(\mathcal{G},\mathbb{C}^n)$, respectively.

Moreover,

$$F_1 y := F_{11}y(a) + F_{12}y(b) ,$$

$$H_1 y := H_{11}y(a) + H_{12}y(b)$$

furnish linear mappings of $\mathcal{U}_o$ into $\mathbb{C}^n$. Then

$$F := (F_o,F_1), \quad H := (H_o,H_1)$$

yield linear mappings of $\mathcal{U}_o$ into

$$\mathcal{R} := C_o(i,\mathbb{C}^n) \times \mathbb{C}^n \quad \text{or}$$

$$\mathcal{R} := H(\mathcal{G},\mathbb{C}^n) \times \mathbb{C}^n ,$$

respectively. With these definitions and notations one can now consider the boundary-eigenvalue problem (BEVP)

(1) $$Fy + \lambda Hy = 0 \qquad (y \in \mathcal{U}_o) .$$

We assume now that $F_{o1}(x) + \lambda H_{o1}(x)$ be invertible for all $x \in i$ or $x \in \mathcal{G}$, respectively, and for all $\lambda \in \mathbb{C}$. Then the existence and uniqueness theorems for explicit linear differential systems can be applied. Then there exists a unique fundamental matrix $Y(x;\lambda)$ with

$$(F_{o1}(x) + \lambda H_{o1}(x))Y'(x;\lambda) + (F_{o2}(x) + \lambda H_{o2}(x))Y(x;\lambda) = 0 ,$$

$$Y(a;\lambda) = E ,$$

where $'$ denotes the x-derivative and E the (n,n)-unit matrix. Since

$$(F_o + \lambda H_o)y = 0$$

is equivalent to

(2) $$y(x) = Y(x,\lambda)c$$

with $c \in \mathbb{C}^n$, (1) becomes equivalent to (2) and

(3) $$D(\lambda)c = 0$$

with

$$D(\lambda) := (F_{11} + \lambda H_{11}) + (F_{12} + \lambda H_{12})Y(b;\lambda) .$$

We set

$$\Delta(\lambda) := \det D(\lambda) .$$

Due to known theorems on parameter dependence, D and Δ are entire analytic functions with values in the (n,n)-matrices and $\mathbb{C}$, respectively.

We give now

Theorem 31 : The order of the pair $\{F,H\}$ coincides with the order of

$$D(\lambda) = D_o + \lambda D_1 + \lambda^2 D_2 \quad \dots ,$$

that is to say with the order of 0 as a zero of Δ.

Proof : Let $p \in \mathbb{N}_o$. The power series expansion

$$Y(x;\lambda)\left(c_o + \lambda c_1 + \dots + \lambda^p c_p\right) = g_o(x) + \lambda g_1(x) + \dots + \lambda^p g_p(x) + \mathcal{O}(\lambda^{p+1})$$

and

$$g_\nu(a) = c_\nu \qquad (\nu=0,1,\dots,p)$$

obviously give a bijective linear relation between $c_o,c_1,\dots,c_p \in \mathbb{C}^n$ and some $g_o,g_1,\dots,g_p \in \mathcal{U}_o$. If the c_ν are such that

(*) $$D(\lambda)\left(c_o + \lambda c_1 + \dots + \lambda^p c_p\right) = \mathcal{O}(\lambda^{p+1}) ,$$

then it is evident according to the definition of D that

(**) $$(F_1 + \lambda H_1)\left(g_o + \lambda g_1 + \dots + \lambda^p g_p\right) = \mathcal{O}(\lambda^{p+1}) ,$$

while the definition of Y gives

($\overset{*}{**}$) $$(F_o + \lambda H_o)\left(g_o + \lambda g_1 + \dots + \lambda^p g_p\right) = \mathcal{O}(\lambda^{p+1}) ,$$

or, in combination,

($\overset{**}{**}$) $$(F + \lambda H)\left(g_o + \lambda g_1 + \dots + \lambda^p g_p\right) = \mathcal{O}(\lambda^{p+1}) .$$

Conversely, if ($\overset{**}{**}$) holds, then the right member is explicitly $\lambda^{p+1} Hg_p$; analogous results hold in (**),($\overset{*}{**}$). If then $y_o(x;\lambda)$ is the solution of

$$(F_o + \lambda H_o)y_o = H_o g_p , \quad y_o(a) = 0 ,$$

then ($\overset{*}{**}$) obviously is equivalent to

$$g_o(x) + \lambda g_1(x) + \dots + \lambda^p g_p(x) = Y(x,\lambda)\left(c_o + \lambda c_1 + \dots + \lambda^p c_p\right) + \lambda^{p+1} y_o(x;\lambda)$$

with $c_\nu = g_\nu(a)$ $(\nu=0,1,\dots,p)$ and $\lambda^{p+1} y_o(x;\lambda) = \mathcal{O}(\lambda^{p+1})$.

But then (*) is recovered from (**). - Consequently the dimensions of the spaces $\mathcal{E}_p$ for $\{F,H\}$ and those for $\{D_o,D_1,D_2,\dots\}$ agree. In particular, the first statement of the theorem holds. The second one follows with Theorem 30 .

We remark that Theorem 31 can be considerably generalized: On the one hand, instead of the "boundary values" at a and b one can consider a more general functional; on the other hand, λ in the differential equation and functional part may occur in an arbitrary holomorphic way. The precise formulation with the necessary functionalanalytic aspects is some what more complicated; therefore we have here restricted ourselves to the present case which is tailored to the intended applications.

Assume now that also for F^* with $F^*_{o1}, F^*_{o2}, F^*_{11}, F^*_{12}$ and for H^* with H^*_{o1}, $H^*_{o2}, H^*_{11}, H^*_{12}$ the presuppositions hold which have just been made for F and H ; moreover, S with $S_{o1}, S_{o2}, S_{11}, S_{12}$ and S^* with $S^*_{o1}, S^*_{o2}, S^*_{11}, S^*_{12}$ shall satisfy the presuppositions made for F , but without the invertibility assumption for F_{o1} . (Here the $*$ will not denote the adjoint matrix!)

The transposition sign ^{T} shall transform a column vector into a row vector, an (n,n)-matrix into the transposed matrix.

Now we introduce for $f = (f_o, f_1) \in \mathcal{R}$, $g = (g_o, g_1) \in \mathcal{R}$ the "scalar product"

$$[f,g] := \int_a^b g_o(x)^T \cdot f_o(x)dx + g_1^T \cdot f_1 \quad .$$

For fixed subspaces $\mathcal{U}, \mathcal{U}^*$ with

$$\mathcal{U}_o \supset \mathcal{U} \supset F_o^{-1} H_o \mathcal{U}_o \ ,$$
$$\mathcal{U}_o \supset \mathcal{U}^* \supset F_o^{*-1} H_o^* \mathcal{U}_o \ ,$$

there shall hold

$$\left.\begin{array}{l}[Fu, S^*v] = [Su, F^*v] \\ [Hu, S^*v] = [Su, H^*v]\end{array}\right\} \qquad (u \in \mathcal{U},\ v \in \mathcal{U}^*) \ .$$

Moreover, we require that the two $(2n,2n)$-matrices

$$\begin{pmatrix} F_{11} + \lambda H_{11}, & F_{12} + \lambda H_{12} \\ S_{11} & S_{12} \end{pmatrix} \quad , \quad \begin{pmatrix} F^*_{11} + \lambda H^*_{11}, & F^*_{12} + \lambda H^*_{12} \\ S^*_{11} & S^*_{12} \end{pmatrix}$$

have rank $2n$ for all $\lambda \in \mathbb{C}$.

Let finally $Y^*(x;\lambda)$ be the fundamental matrix to

$$(F_o^* + \lambda H_o^*)Y^* = 0 \quad \text{with} \quad Y^*(a;\lambda) = E$$

and

$$D^*(\lambda) := (F^*_{11} + \lambda H^*_{11}) + (F^*_{12} + \lambda H^*_{12})Y^*(b;\lambda) \ .$$

Then we show

Theorem 32 : The multiplicity as well as the order are the same for $D(\lambda)$ and $D^*(\lambda)$. Therefore the multiplicity as well as the order for $\{F,H\}$ and $\{F^*,H^*\}$ are the same.

Proof : From

$$(F_o + \lambda H_o)y = 0 \ , \ (F_o^* + \lambda H_o^*)y^* = 0$$

follows due to our presuppositions

$$(S_{11}^* y^*(a) + S_{12}^* y^*(b))^T((F_{11} + \lambda H_{11})y(a) + (F_{12} + \lambda H_{12})y(b))$$

$$= ((F_{11}^* + \lambda H_{11}^*)y^*(a) + (F_{12}^* + \lambda H_{12}^*)y^*(b))^T(S_{11}y(a) + S_{12}y(b)).$$

With the definitions

$$\Sigma(\lambda) := S_{11} + S_{12}Y(b;\lambda) ,$$

$$\Sigma^*(\lambda) := S_{11}^* + S_{12}^* Y^*(b;\lambda) ,$$

and with $y = Yc, y^* = Y^*d$ one obtains the matrix identity

(x) $$\Sigma^*(\lambda)^T D(\lambda) = D^*(\lambda)^T \Sigma(\lambda) .$$

If now

$$D(\lambda)\left(c_o + \lambda c_1 + \ldots + \lambda^p c_p\right) = \mathcal{O}(\lambda^{p+1}) ,$$

in particular $D(0)c_o = 0$, and if one expands

$$\Sigma(\lambda)\left(c_o + \lambda c_1 + \ldots + \lambda^p c_p\right) = d_o + \lambda d_1 + \ldots ,$$

one obtains with (x)

$$D^*(\lambda)^T\left(d_o + \lambda d_1 + \ldots + \lambda^p d_p\right) = \mathcal{O}(\lambda^{p+1}).$$

If $c_o \neq 0$ one has also

$$\Sigma(0)c_o = d_o \neq 0 ;$$

for $D(0)c_o = 0$ implies that $(c_o, Y(b;0)c_o) \neq 0$ annihilates the first n rows of the matrix

$$\begin{pmatrix} F_{11} & F_{12} \\ S_{11} & S_{12} \end{pmatrix} ,$$

therefore not the remaining rows because of the assumption on the rank. Therefore one has

$$\dim \mathcal{L}_p \leqq \dim \mathcal{L}_p^{*T}$$

with the spaces $\mathcal{L}$ for

$$D^{*T}(\lambda) = D_o^{*T} + \lambda D_1^{*T} + \ldots$$

in the right member. In particular one has therefore in obvious notation

$$\text{mult}_o(D(\lambda)) \leqq \text{mult}_o(D^{*T}(\lambda)) ,$$

$$\text{ord}_o(D(\lambda)) \leqq \text{ord}_o(D^{*T}(\lambda)) .$$

Now everything holds in the reversed order for $D \longleftrightarrow D^*$. On the other hand one has

$$\text{mult}_o(D^{*T}(\lambda)) = \dim \ker D^{*T}(0) =$$

$$= \dim \ker (D^*(0)) = \text{mult}_o(D^*(\lambda))$$

and

$$\text{ord}_o(D^{*T}(\lambda)) = \text{ord}_o \det D^{*T}(\lambda) =$$

$$= \text{ord}_o \det D^*(\lambda) = \text{ord}_o(D^*(\lambda)) .$$

This yields the assertions if for the last statement one calls upon Theorem 31.

If F,F^* are inhomogeneous linear functions of $\mu \in \mathbb{C}$, that is if one considers correspondingly $F + \mu G, F^* + \mu^* G^*$ in place of F,F^*, and if the presuppositions on F,F^* are now satisfied for all $\mu \in \mathbb{C}$, in particular the adjointness in fixed $\mathcal{U}, \mathcal{U}^*$, the invertibility of $F_{01} + \mu G_{01} + \lambda H_{01}$, $F^*_{01} + \mu G^*_{01} + \lambda H^*_{01}$ and the rank condition for the boundary matrices, then one obviously has realized (0) with $\mathcal{R}^* = \mathcal{R}$. Since one can further consider $F + \lambda_o H$, $F^* + \lambda_o H^*$ in place of F,F^* and $\lambda - \lambda_o$ in place of λ, our theorems can be summarized in:

Theorem 33 : For the considered boundary-eigenvalue problems

$$Fy + \lambda Hy + \mu Gy = 0 \qquad (y \in \mathcal{U}),$$

$$F^*y^* + \lambda H^*y^* + \mu G^*y^* = 0 \qquad (y^* \in \mathcal{U}^*),$$

a suitable function with the properties (2) and (7) is

$$\Delta(\lambda,\mu) := \det D(\lambda,\mu) .$$

1.1.10. Application to Hill's differential equation in the real domain.

For Hill's differential equation

(1) $$y''(x) + (\lambda + \mu\Theta(x))y(x) = 0 \qquad (x \in \mathbb{R})$$

with 2π - periodic continuous complex valued Θ, we consider with arbitrary $\nu \in \mathbb{C}$ the characteristic eigenvalue problem

(2) $$y(x + 2\pi) = e^{2\pi i \nu} y(x) .$$

$\mathcal{U}$ and $\mathcal{R}$ can be chosen as the complex valued twice continuous differentiable or continuous functions, respectively, on $\mathbb{R}$ with (2). $\mathcal{U}^*$ and $\mathcal{R}^*$ are chosen in an analogous manner with

(2*) $$y(x + 2\pi) = e^{-2\pi i \nu} y(x) ,$$

F and F^* are chosen as the second x-derivation, H and H^* as the identity (embedding), and G and G^* as multiplication by Θ. One also chooses S and S^* in $\mathcal{U}$ and $\mathcal{U}^*$, respectively, as the identity (embedding) and for $f \in \mathcal{R}, g \in \mathcal{R}^*$ one defines

(3) $$[f,g] := \frac{1}{2\pi} \int_a^{a+2\pi} f(x)g(x)dx ,$$

which is independent of $a \in \mathbb{R}$, the integrand being 2π - periodic. Then 1.1.1., (0) is satisfied in the usual manner. (1) is immediately clear. For (2) one chooses for instance

(4) $$\Delta(\lambda,\mu) := \det(Y(2\pi;\lambda,\mu) - e^{2\pi i \nu} E) =$$

$$= 2e^{2\pi i \nu}\left(\cos 2\nu\pi - \frac{1}{2}(y_I(2\pi;\lambda,\mu) + y'_{II}(2\pi;\lambda,\mu))\right)$$

where

$$Y = \begin{pmatrix} y_I & , & y_{II} \\ y_I' & , & y_{II}' \end{pmatrix}$$

is the fundamental matrix with $Y(0;\lambda,\mu) = E$. (3) holds with - here and in the following numbered with $\mathbb{Z}$ instead of $\mathbb{N}$ -

$$\lambda_n = (\nu + n)^2 \qquad (n \in \mathbb{Z})$$

and

$$y_n^*(x) = e^{-i(\nu+n)x} .$$

For $2\nu \notin \mathbb{Z}$ all the λ_n are different. If ν is an integer, $\nu = 0$ without less of generality, then one has

$$\lambda_n = \lambda_{-n} \qquad (n \in \mathbb{N}) .$$

If ν is one half of an odd integer, $\nu = \frac{1}{2}$ without less of generality, then one has

$$\lambda_n = \lambda_{-n-1} \qquad (n \in \mathbb{N}_0) .$$

(3) is satisfied. For (4) one chooses for the number sequences (now numbered with $\mathbb{Z}$)

$$\| \ \|_2 = | \ |_2 , \| \ \|_1 = | \ |_1 , \| \ \|_\infty = | \ |_\infty ,$$

that is

$$\eta = \delta = (1)_{n \in \mathbb{Z}} .$$

Then

$$\|f\| := \left(\frac{1}{2\pi} \int_a^{a+2\pi} \left|f(x)e^{-i\nu x}\right|^2 dx \right)^{\frac{1}{2}} ,$$

$$\lceil f \rceil := \frac{1}{2\pi} \int_a^{a+2\pi} |f(x)e^{-i\nu x}| dx ,$$

$$|f| = |f|_o := \max\{|f(x)e^{-i\nu x}| : x \in \mathbb{R}\}$$

become norms in $\mathcal{R}$. (4.5) holds as Parseval's equation for the Fourier coefficients of $f(x)e^{-i\nu x}$, (4.6) is a rough estimate of the Fourier coefficients, and (4.7) is a simple series estimate which is non-trivial only if the Fourier series of $f(x)e^{-i\nu x}$ is absolutely convergent in the norm $| \ |$.

In the following

$$\max\{|\Theta(x)| : x \in \mathbb{R}\} =: \hat{\gamma}$$

is to be used. For (5) one takes $\gamma^{(1)} = (1)_{n \in \mathbb{Z}}$ and $\gamma^{(2)} = \gamma = (\hat{\gamma})_{n \in \mathbb{Z}}$, further $A_k = id_{\mathcal{R}}$ $(k \in \mathbb{N})$. B_k $(k \in \mathbb{N})$ is explained in $\mathcal{R}$ as multiplication by Θ. With Lemma II it is easily recognized that $B_k \in (II;1,\gamma^{(2)}) \cap (II;2,\gamma^{(2)})$. With

$$\zeta_1 = 1, \ \zeta_k = 0 \qquad (k \geq 2)$$

one has (5) a),b),c). Then one has in (5.3)

$$\hat{M} = \infty .$$

It is wellknown that (6) is satisfied.

In the theorems on the eigenvalues we have in 1.1.3. $\hat{L} = \infty$ and always

$$r_n = \frac{1}{2} d_n \hat{\gamma}^{-1} ,$$

and in every case

$$r_n \to \infty \qquad (n \to \pm\infty) ,$$

and, of course , $r_{n/|n|}$ is bounded from below and above if $\hat{\gamma} > 0$.

Therefore with Theorem 10, Theorem 11 power series expansion, estimate and asymptotics of the $\lambda_n(\mu)$ can be obtained if $2\nu \notin \mathbb{Z}$.

However, if $2\nu \in \mathbb{Z}$, then around those λ_n which are double zeros of $\Delta(\cdot,0)$ there can occur as solutions either two power series in μ or one power series in $\mu^{1/2}$. Then one can always consider the symmetric functions of the two branches for $|\mu| < r_n$ which are there holomorphic functions and obtain similar estimates (see Theorem 5, Therorem 7).

If Θ is an even function of x then one can separate into even and odd solutions and thus arrive at the case of simple zeros. This situation has been discussed at great length in MS for the special case of Mathieu functions. Therefore one has here always a power series solution with Theorem 10, Theorem 11. Also for the case of a real function Θ and $\nu \in \mathbb{R}$ one can say that always a power series solution in μ occurs. In this case the problems for real μ are self-adjoint and consequently lead for real μ to real $\lambda_n(\mu)$; therefore odd powers of $\mu^{1/2}$ cannot occur in an expansion of $\lambda_n(\mu)$.

We abstain from giving more details.

Now we get to the further presuppositions in 1.1.4.. For (7) one shows at first from (4)

$$\Delta^*(\lambda,\mu) = e^{-4\pi i\nu}\Delta(\lambda,\mu)$$

and with that 1.1.9., Theorem 31 yields the remaining statements by transcribing to a (2,2) system; it is possible to apply 1.1.9., Theorem 32, but not necessary because of the simple deduction from (4). (8) is satisfied in known manner and in connexion with it one can choose $A = G$ in c) and e). (9), (10), (11) are evident.

Now Theorem 21 can be applied, say with $B = A = id_{\mathcal{R}}$. With this choice one can draw upon Theorem 29 and has in any case with $A = B = id_{\mathcal{R}}$

$$D_m = \mathcal{O}(m^{-1}) .$$

This gives for every L - integrable function the following result: The series after eigen solutions and - should the case arise - principal solutions which originate for $\mu \neq 0$ are equiconvergent with the modified Fourier series for $\mu = 0$, with respect to

$$\sum_{m=-n-1}^{n} \qquad (n \to \infty)$$

in the case $\nu = \frac{1}{2}$, with respect to

$$\sum_{m=-n}^{n} \qquad (n \to \infty)$$

in the case $\nu = 0$ and in all other cases with $\nu \notin \mathbb{Z}$. It is after all essential that those terms in the series expansion which belong to the same eigenvalue at $\mu = 0$, must not be separated.

The asymptotics of eigen and principal solutions for large eigenvalue numbers follows in usual manner from Theorem 27, Theorem 28 with $A = B = id_{\mathcal{R}}$ and

$$c_m = \mathcal{O}(m^{-1}) \ .$$

We resign the detailed formulation and the explicit explanation of the elementary estimates. With respect to that reference is made to F.W.Schäfke (1960).

1.1.11. Application to Hill's differential equation in the complex domain.

Let

$$-\infty \leqq s_1 < s_{1o} < s_{2o} < s_2 \leqq +\infty$$

and

$$\mathbb{C} \supset \{z : s_1 < \mathrm{Im}\, z < s_2\} =: \mathcal{T} \supset \{z : s_{1o} < \mathrm{Im}\, z < s_{2o}\} =: \mathcal{T}_o \ .$$

With a

$$\Theta : \mathcal{T} \longrightarrow \mathbb{C}$$

which is holomorphic and 2π - periodic, we consider Hill's differential equation

(1) $$y''(x) + (\lambda + \mu\Theta(x))y(x) = 0 \qquad (x \in \mathcal{T})$$

and with arbitrary $\nu \in \mathbb{C}$ the characteristic eigenvalue problem

(2) $$y(x + 2\pi) = e^{2\pi i\nu}y(x)$$

and adjoint to it

(2*) $$y^*(x + 2\pi) = e^{-2\pi i\nu}y^*(x) \ .$$

For the application of our theory s_{1o}, s_{2o} are fixed and $\mathcal{R}$ is chosen as the space of the holomorphic functions in $\mathcal{T}_o$ which satisfy (2) and which are still continuous in $\overline{\mathcal{T}}_o$; analogously $\mathcal{R}^*$ with (2*) . $\mathcal{U}$ is chosen as the space of the functions which are holomorphic in $\mathcal{T}_o$, which satisfy (2), and which are together with their first and second derivatives still continuous in $\overline{\mathcal{T}}_o$; analogously $\mathcal{U}^*$ with (2*).

Of course, F and F^* are taken as the second derivative, H and H^* as the identity (embedding), and G and G^* as multiplication by Θ . Also S and S^* are taken as the identity (embedding), and for $f \in \mathcal{R}$, $g \in \mathcal{R}^*$ we define

(3) $$[f,g] := \frac{1}{2\pi} \int_a^{a+2\pi} f(x)g(x)dx \ ,$$

which is independent of $a \in \bar{\mathcal{T}}_o$ because of the periodicity of the integrand. With that (0) and (1) are satisfied. For (2) an $a_o \in \mathcal{T}_o$ is fixed and one sets again

$$(4) \quad \begin{cases} \Delta(\lambda,\mu) = \det\left(Y(a_o + 2\pi;\lambda,\mu) - e^{2\pi i \nu}E\right) = \\ = 2e^{2\pi i \nu}\left(\cos 2\nu\pi - \frac{1}{2}(y_I(a_o + 2\pi;\lambda,\mu) + y'_{II}(a_o + 2\pi;\lambda,\mu))\right) , \end{cases}$$

where

$$Y = \begin{pmatrix} y_I & y_{II} \\ y'_I & y'_{II} \end{pmatrix}$$

is the fundamental matrix with $Y(a_o;\lambda,\mu) = E$. For (3) there is, as in 1.1.10., with the numbering $\mathbb{Z}$ instead of $\mathbb{N}$,

$$\lambda_n = (\nu+n)^2$$

and in line with this

$$y_n^*(x) = e^{-i(\nu+n)x} .$$

As above, we restrict ourselves to $\nu = 0$ or $\nu = \frac{1}{2}$ if $2\nu \in \mathbb{Z}$.

For (4) it is expedient to choose

$$||f||^2 := \max\left\{\frac{1}{2\pi}\int_a^{a+2\pi} |f(x)e^{-i\nu x}|^2 dx : a \in \bar{\mathcal{T}}_o\right\} ,$$

$$\boxed{f} := \max \frac{1}{2\pi}\left\{\int_a^{a+2\pi} |f(x)e^{-i\nu x}| dx : a \in \bar{\mathcal{T}}_o\right\} ,$$

$$|f| = |f|_o := \max\left\{|f(x)e^{-i\nu x}| : x \in \bar{\mathcal{T}}_o\right\} ,$$

with the integrals being taken along the straight line connecting a and $a + 2\pi$. Correspondingly one chooses for $\alpha \in \mathbb{C}^{\mathbb{Z}}$

$$||\alpha||_2^2 := \max\left\{\sum_{n\in\mathbb{Z}} |\alpha_n|^2 e^{-2sn} : s_{1o} \leq s \leq s_{2o}\right\} ,$$

$$||\alpha||_1 := \sup\left\{\sum_{n\in\mathbb{Z}} |\alpha_n| e^{-sn} : s_{1o} \leq s \leq s_{2o}\right\} ,$$

$$||\alpha||_\infty := \sup\left\{|\alpha_n| e^{-sn} : s_{1o} \leq s \leq s_{2o} , n \in \mathbb{Z}\right\} .$$

With

$$\delta = \eta := (1)_{n\in\mathbb{Z}}$$

(4.1), (4.2), (4.3) hold. (4.5) originates from Parseval's equation for the real Fourier series which arise on $\operatorname{Im} z = c$, (4.6) from the pertinent coefficient estimate, and (4.7) again from the pertinent series estimate. With that (4) holds; $||\ ||$, $\boxed{\ }$ and $|\ | = |\ |_o$ are even norms in $\mathcal{R}$.

For (5) one uses of course

$$\hat{\gamma} := \max\{|\Theta(x)| : x \in \overline{\mathcal{T}}_o\}$$

and sets, as in 1.1.10.,

$$\gamma^{(1)} := (1)_{n\in\mathbb{Z}}\ ,\ \gamma^{(2)} = \gamma := (\hat{\gamma})_{n\in\mathbb{Z}}\ ,$$

$$A_k = id_{\mathcal{R}} \qquad (k \in \mathbb{N})$$

and defines B_k in $\mathcal{R}$ as multiplication by Θ . Then with

$$\zeta_1 = 1,\ \zeta_k = 0 \qquad (k \geqq 2)$$

(5) a) b) c) are satisfied. For that one can again refer to Lemma II in 1.1.10..

The conclusions for the eigenvalues correspond to those in 1.1.10. . In the respective estimates $\mathcal{T}_o$ should be so chosen that $\hat{\gamma}$ is small.

As was the case in 1.1.10., the verification of (7) to (11) is easy.

We now apply again Theorem 21 with $B = A = id_{\mathcal{R}}$ and draw upon Theorem 29. Here one has again

$$D_m = \mathcal{O}(m^{-1}) \ .$$

In this way one obtains at least for $f \in \mathcal{R}$ the equiconvergence statements corresponding to those in 1.1.10. .

For that goal one can now take into account that every f , which is holomorphic in $\mathcal{T}$ and has the property (2), can be represented with a g which is holomorphic in $\mathcal{T}$ and has the property (2) in the form

$$f = R(\hat{\lambda},0)g$$

provided $\hat{\lambda}$ is not eigenvalue to $\mu = 0$. One has only to set

$$g(x) := f''(x) + \hat{\lambda}f(x) \ .$$

Then one can apply Theorem 29 with $A = id_{\mathcal{R}}$ and

$$B = R(\hat{\lambda},0),\ \beta = \left(|\lambda_n - \hat{\lambda}|^{-1}\right)_{n\in\mathbb{Z}}$$

and recognizes here easily with the possible restriction to $\hat{\gamma} > 0$ that

$$\sum_m D_m < \infty \ .$$

If finally one varies $\mathcal{T}_o$, one obtains the result that every function which is holomorphic in $\mathcal{T}$ and has the property (2) can be expanded in a locally uniformly and absolutely convergent series in terms of the eigen solutions and the principal solutions, respectively; but for $2\nu \in \mathbb{Z}$ it is necessary to unite two terms which belong to the same unperturbed eigenvalue into one term of the series.

With respect to the asymptotics of eigen and principal solutions for large numbers of the eigenvalue the results are analogous to those in 1.1.10.. Just as we did there, we resign here the detailed formulation and the explicit explanation of the elementary estimates.

It is scarcely necessary to call attention to the possibility of splitting into even and odd functions in case of an even function Θ and a stripe which is symmetric to the real axis.

1.1.12. Application to the spheroidal differential equation in the real domain.

We treat - like in MS 3.2. - the eigenvalue problems of the spheroidal functions

$$ps_n^m(x;\gamma^2)$$

of integer order and integer degree:

$$m = 0,1,2,\dots ,$$

$$n = m,\ m+1,\ m+2,\ \dots .$$

We proceed, however, otherwise than in MS and obtain with two different applications of our theory on the one hand essentially the results in MS , but on the other hand new results.

For a given

$$m = 0,1,2,\dots$$

the spheroidal differential equation in the form

(1) $$[(1-x^2)y'(x)]' + \left[\lambda + \gamma^2\left(\frac{1}{2} - x^2\right) + \frac{-m^2}{1-x^2}\right]y(x) = 0$$

and the eigenvalue condition

(2) $$y \text{ continuous in } -1 \leqq x \leqq 1$$

are considered. A splitting into even and odd functions, although possible , to be treated in an analogous way as formerly, will not be made.

We choose $\mathcal{R} = \mathcal{R}^*$ as the space of the continuous complex valued functions on $-1 \leqq x \leqq 1$ and

$$[f,g] := \frac{1}{2}\int_{-1}^{1} f(x)g(x)dx .$$

$\mathcal{U} = \mathcal{U}^*$ shall be the subspace of the continuous functions f on $-1 \leqq x \leqq 1$ with two continuous derivatives in the open interval and with the additional properties that

$$(1-x^2)f'(x)$$

vanishes for $x \to +1$ and $x \to -1$ and that

(3) $$\left[(1-x^2)f'(x)\right]' - \frac{m^2}{1-x^2}f(x)$$

has finite limits for $x \to +1$ and $x \to -1$. (For $m \in \mathbb{N}$ there follows, by the way, $f(x) \to 0$ as $x \to \pm 1$).

Of course, $H = H^* = S = S^*$ are now chosen as the identity (embedding) and $F = F^*$ as the mapping given with (3). $G = G^*$ becomes the multiplication by $(\frac{1}{2} - x^2)$.

Then 1.1.1. ⓪ , ① hold: The selfadjointness is immediately checked with the requirements on $\mathcal{U} = \mathcal{U}^*$. Of course, EP and aEP correspond just to (1), (2).

Let now y_I be the solution of (1) with

$$y_I(x) = (1-x)^{m/2}\,\mathcal{P}(1-x)\ ,\ \mathcal{P}(0) = 1\ ,$$

where $\mathcal{P}$ is a power series which converges at least in $|1-x| < 2$, and is an entire function of λ,γ^2 - see MS 3.12. . Then we define

$$\Delta(\lambda,\gamma^2) := y_I(0;\lambda,\gamma^2)\,y_I'(0;\lambda,\gamma^2)\quad ,$$

with which obviously (2) holds. For (3) the eigenvalues

$$\lambda_n := (m+n-1)(m+n) \qquad (n \in \mathbb{N})$$

are simple zeros of $\Delta(\cdot,0)$. This has been shown already in MS 3.21. , but it results now, however, more generally from the considerations which are given below for the proof of (7). The eigensolutions to λ_n are

$$y_n(x) := y_n^*(x) := \sqrt[+]{(2m+2n-1)\ \frac{(n-1)!}{(2m+n-1)!}}\ P^m_{m+n-1}(x)\ .$$

For (4) we set in each case

$$\|\alpha\|_2 := |\alpha|_2$$

and

$$\|f\|^2 := \frac{1}{2}\int_{-1}^{1} |f(x)|^2 dx$$

with which (4.1) and (4.5) hold. $\|\ \|$ is even a norm in $\mathcal{R}$.

We now note - according to MS 3.21. - that

$$\max\{|y_n(x)|\ :\ -1 \leqq x \leqq 1\} \leqq M_m\, n^{1/2} \qquad (n \in \mathbb{N})$$

with a constant $M_m > 0$ which depends on m only. On the other hand one has for $0 < \varepsilon < 1$

$$\max\{|y_n(x)|\ :\ -1+\varepsilon \leqq x \leqq 1-\varepsilon\} \leqq N_m(\varepsilon) \qquad (n \in \mathbb{N})$$

with a constant $N_m(\varepsilon)$ which depends on m and ε only.

Thus one has the following interesting possibilities in (4).

On the one hand one can choose

$$\overline{|f|} := \|f\|\ ,$$

$$\|\alpha\|_\infty := |\alpha|_\infty\ ,$$

$$\delta = (1)_{n\in\mathbb{N}}$$

and

$$|f|_0 := \max\{|f(x)|\ :\ -1 \leqq x \leqq 1\}\ ,$$

$$\eta := \left(M_m n^{1/2}\right)_{n\in\mathbb{N}}\ ,$$

$$\|\alpha\|_1 := |\alpha\eta|_1\ .$$

Then (4.2), (4.3), (4.6), (4.7) are satisfied. (4.7) is, of course, a simple series estimate.

Another choice is

$$\overline{|f|} := \frac{1}{2} \int_{-1}^{1} |f(x)|\,dx \quad ,$$

$$\|\alpha\|_\infty := |\alpha\delta^{-1}|_\infty \quad ,$$

$$\delta := \left(M_m n^{1/2}\right)_{n \in \mathbb{N}}$$

and with a fixed ε

$$|f|_0 := \frac{1}{N_m(\varepsilon)} \max\{|f(x)| \; : \; -1 + \varepsilon \leqq x \leqq 1 - \varepsilon\} \quad ,$$

$$\eta := (1)_{n \in \mathbb{N}}$$

$$\|\alpha\|_1 := |\alpha|_1 \quad .$$

Also with these choices (4.2), (4.3), (4.6), (4.7) hold. (4.6) is the elementary rough coefficient estimate and (4.7) is again a simple series estimate.

For (5) one proceeds similarly as in 1.1.10. and obtains

$$\gamma^{(1)} := (1), \; \gamma^{(2)} := \gamma := \left(\frac{1}{2}\right) \quad .$$

Now we get to (6). One sets for $f \in \mathcal{R}$, $-1 < x < 1$

$$u(x) := \int_{-1}^{x} y_I(-\xi) f(\xi)\,d\xi \cdot y_I(x) + \int_{x}^{1} y_I(\xi) f(\xi)\,d\xi \cdot y_I(-x)$$

and uses

$$u'(x) = \int_{-1}^{x} \cdots d\xi \cdot y_I'(x) - \int_{x}^{1} \cdots d\xi \cdot y_I'(-x) \quad ,$$

$$y_I(-x) y_I'(x) + y_I(x) y_I'(-x) = \frac{2\Delta}{1-x^2} \quad ,$$

$$u''(x) = \frac{2\Delta}{1-x^2} f(x) + \int_{-1}^{x} \cdots d\xi \cdot y_I''(x) + \int_{x}^{1} \cdots d\xi \cdot y_I''(-x) \quad .$$

Thus one finds

$$u \in \mathcal{U}, \; (f + \lambda H + \mu G) u = 2\Delta f \quad .$$

Of course, the behavior of $y_I(x)$ at $x = -1$ and correspondingly of $y_I(-x)$ at $x = 1$ has to be taken care of; see MS 3.12. .

Then (6) holds with

$$z(x) := \frac{1}{2\Delta} u(x) \quad .$$

In order to verify (7) we introduce for $\kappa = 0,1,2,\ldots$ the notation

$$y_{[\kappa]}(x;\lambda,\gamma^2) := \frac{1}{\kappa!} \left(\frac{\partial}{\partial\lambda}\right)^{\kappa} y_I(x;\lambda,\gamma^2)$$

and $y_{[-1]} := 0$. Then for $\kappa = 0,1,2,\ldots$ there obviously holds

$$\left[(1-x^2) y_{[\kappa]}'(x)\right]' + \left[\lambda + \gamma^2\left(\frac{1}{2} - x^2\right) + \frac{-m^2}{1-x^2}\right] y_{[\kappa]}(x) = -y_{[\kappa-1]}(x) \quad .$$

If now λ is precisely a k - fold zero of $\Delta(\cdot,\gamma^2)$, then one obtains easily that either

$$y'_{[o]}(0) = y'_{[1]}(0) = \ldots = y'_{[k-1]}(0) = 0,\ y'_{[k]}(0) \neq 0$$

or

$$y_{[o]}(0) = y_{[1]}(0) = \ldots = y_{[k-1]}(0) = 0,\ y_{[k]}(0) \neq 0 \quad ,$$

and conversely. In this case one has

$$y_{[o]},\ y_{[1]},\ \ldots,\ y_{[k-1]} \in \mathfrak{U},\ y_{[k]} \notin \mathfrak{U}$$

and these functions are in the first case even, in the second case odd. This provides k linearly independent principal vectors. But one still has to show that there are not more of them. For that purpose let $y_{[k-1]} \in \mathfrak{U}$ with a $\kappa = 0,1,2,\ldots$ and $h \in \mathfrak{U}$ such that

$$(F + \lambda H + \gamma^2 G)h = -y_{[\kappa-1]} \quad ,$$

Then

$$h = y_{[\kappa]} + \beta y_{[o]}$$

with $\beta \in \mathbb{C}$ because $h - y_{[\kappa]}$ is a solution of the homogeneous differential equation and belongs at $x = 1$ to the index $+\frac{m}{2}$; therefore in this case $y_{[\kappa]} \in \mathfrak{U}$. Now this consideration demonstrates: If λ is a k - fold zero of $\Delta(\cdot,\gamma^2)$, then $y_{[o]},\ldots,y_{[k-1]}$ form a basis of the principal space. This proves (7) .

For (8) one chooses, of course,

$$|f| := \max\{|f(x)| : -1 \leq x \leq 1\} \quad .$$

Then, as the preceding construction shows, b) is satisfied, equally c), for instance with $A := G$. d) and e) hold because even in $\mathfrak{R}$ one always has $\|f\| \leq |f|$.

(9) and (10) are obviously true. Of course, (11) holds only with the first choice of $|f|_o$.

We can now pass over the consequences of 1.1.3. as being known: of course one has the stronger estimates available when splitting the eigenvalue problem into the two eigenvalue problems of the even and the odd functions.

The consequences of 1.1.6., Theorem 21 with $A = B = id_{\mathfrak{R}}$ are of interest.

It yields with the first choice for (4) the equiconvergence of the series in terms of eigenfunctions and, if γ^2 is an exceptional value, of the principal solutions with series expansion in terms of Legendre functions for every square integrable function, uniformly on $-1 \leq x \leq 1$.

With the second choice for (4) the equiconvergence of those series results for every integrable function uniformly on every subinterval $-1 + \varepsilon \leq x \leq 1 - \varepsilon$.

This yields essential additions to MS 3.23. , Theorem 4 in two directions.

Again the importance of the estimates in 1.1.8. for the asymptotics at $n \to \infty$ needs no further comment.

1.1.13. Application to the spheroidal differential equation in the complex domain.

Here for the spheroidal differential equation

(1) $$[(1-z^2)y'(z)]' + \left[\lambda+\gamma^2(1-z^2) + \frac{-\mu^2}{1-z^2}\right] y(z) = 0$$

the characteristic eigenvalue problem

(2) $$y(ze^{\pi i}) = e^{-\pi i(\nu+1)}y(z)$$

between two confocal ellipses with the focal points +1, -1 is of interest. In MS 3.51., 3.544. a direct treatment has been given, although only for $\nu \not\equiv \frac{1}{2} (\mathrm{mod}\ 1)$, and for pertinent normal γ^2 an expansion theorem and asymptotic formulas have been derived.

Here we will only point out that these results together with the supplements for $\nu + \frac{1}{2} \in \mathbb{Z}$ and for exceptional values γ^2 can be deduced from the results of 1.1.11. for Hill's differential equation by a simple transformation.
In fact, if one applies to (1) the transformation

(3) $$z = \cos\tau\ ,\ y(z) = (z^2-1)^{-1/4}g(\tau)$$

one obtains - see MS 3.14., (22), (23) -

(4) $$g''(\tau) + \left(\lambda + \frac{1}{4} + \gamma^2 \sin^2\tau + \frac{\frac{1}{4} - \mu^2}{\sin^2\tau}\right) g(\tau) = 0$$

and the eigenvalue problem (2) transforms into

(5) $$g(\tau+\pi) = e^{-\pi i(\nu+1/2)}g(\tau)$$

in a stripe parallel to the real axis within the lower half plane; besides one has to identify $(z^2-1)^{1/2} = i \sin\tau$ and

$$\left(z \pm (z^2-1)^{1/2}\right)^\rho = e^{\pm i\rho\tau} .$$

With that the reduction to 1.1.11. is clear: for instance, one can choose

$$\Theta(x) := \gamma^2\sin^2\frac{x}{2} + \frac{\frac{1}{4} - \mu^2}{\sin^2\frac{x}{2}}\ ,\ x = 2\tau$$

and obtains for the parameter value $\frac{1}{4}$ the desired statements on the expansions. Thus not only the results in MS 3.33. for the Legendre functions and in MS 3.544. for the spheroidal functions are obtained simultaneously. In addition, one also obtains the expansions in terms of eigen and principal solutions including the cases of $\nu + \frac{1}{2} \in \mathbb{Z}$ (in (2)) and of exceptional values γ^2.

For the estimates with respect to γ^2 only, that is with constant μ^2, one will, of course, proceed in analogy to the methods used in MS, either by means of (1), (2) or of (4), (5).

1.2. Simply Separated Operators.

1.2.0. Introduction.

Let elliptical coordinates (ξ,η) be connected with the Cartesian coordinates (x,y) by

$$x \pm iy = c\,\mathrm{Cos}(\xi \pm i\eta)\ ,\quad c > 0\ .$$

The two dimensional (time independent) wave equation $\Delta u + k^2u = 0$ takes in these coordinates (ξ,η) the form

$$Cu := \left(\frac{\partial^2 u}{\partial\xi^2} + 2h^2\,\mathrm{Cos}\,2\xi\cdot u\right) + \left(\frac{\partial^2 u}{\partial\eta^2} - 2h^2\cos 2\eta\cdot u\right) = 0$$

with the abbreviation $h = kc/2$ (see MS 1.135.,(14)). The left member contains a differential operator C which is the sum of two operators, one of them "acting" only on one variable, the other acting only on the other variable.

If, in particular, C is applied to a product

(*) $$u(\xi,\eta) = u_1(\xi)u_2(\eta)\ ,$$

then, with the ordinary differential operators

$$Au_1 := u_1'' + 2h^2\,\mathrm{Cos}\,2\xi\cdot u_1\ ,$$

$$Bu_2 := u_2'' - 2h^2\cos 2\eta\cdot u_2\ ,$$

one obviously has

(**) $$Cu = Au_1\cdot u_2 + u_1\cdot Bu_2\ .$$

We designate such an operator as simply separated.

It is evident that simply separated operators are also obtained if the two dimensional wave equation is separated in cartesian coordinates or in polar coordinates or in parabolic coordinates.

On the other hand, if for instance one considers a solution of the three dimensional wave equation in prolate spheroidal coordinates (ξ,η,φ) in which the φ - dependence is separated off:

$$u(\xi,\eta,\varphi) = v(\xi,\eta)u_3(\varphi)$$

with

$$u_3'' + \mu^2u_3 = 0\ ,$$

then v satisfies the partial differential equation

$$\left(\frac{\partial}{\partial\xi}(1-\xi^2)\frac{\partial v}{\partial\xi} + \gamma^2(1-\xi^2)v - \frac{\mu^2}{1-\xi^2}v\right) +$$

$$+ \left(-\frac{\partial}{\partial\eta}(1-\eta^2)\frac{\partial v}{\partial\eta} - \gamma^2(1-\eta^2)v + \frac{\mu^2}{1-\eta^2}v\right) = 0$$

(see MS 1.133.). One recognizes that it also possesses the simply separated form with a suitable operator C. A similar situation exists for the wave equation in oblate spheroidal coordinates, in spherical coordinates, or in the corresponding parabolic coordinates.

A third possibility of this kind is given with the three dimensional wave equation in spherical coordinates or in conical coordinates after separating off the r - dependence.

Consider now, for instance, in the first mentioned case C as an operator in the space of the entire functions of two variables which are even in each variable, in the first one πi - periodic, in the second one π - periodic, and consider accordingly A and B as operators in the space of the even πi - periodic and π - periodic entire functions, respectively, of one variable. Then it is recognized from (**) that a product of an eigen solution u_1 of A with eigenvalue α and of an eigen solution u_2 of B with eigenvalue β furnishes in (*) an eigen solution of C with eigenvalue $\alpha + \beta$.

If for normal h^2 one uses the corresponding expansion theorem (MS 2.28.) and a suitable integral relation (MS 1.135.) for the evaluation of the coefficients, then one obtains, for instance, a theorem on the expansion of solutions u of $Cu = 0$ in the mentioned space which takes the form

$$u(\xi,\eta) = \sum_{n=0}^{\infty} \gamma_n Ce_{2n}(\xi;h^2)ce_{2n}(\eta;h^2) .$$

This simple situation holds only if h^2 is a normal value. If, however, h^2 is an exceptional value, then for an expansion theorem in one dimension the eigen solutions must be supplemented by the principal solutions - see 1.1. . Then there arises the problem what the corresponding expansion of a u with $Cu = 0$ looks like. Obviously, principal solutions of C to 0 must be introduced which are built up from principal solutions of A and B . There arises the question what the respective Jordan matrices and correspondingly, the grouping into invariant subspaces look like. In other words: how can the normal form of C be constructed from the normal forms of A and B ?

The following investigation serves the purpose of an exact formulation and clarification of these questions.

With respect to that, one has first of all to treat the corresponding algebraic normal form problem for finite dimensional spaces over $\mathbb{C}$. The transition from the functions of one variable (in ξ or η) to the functions of two variables and to (*) is expected to be understood, in an abstract way, algebraically in the tensor product of two spaces (1.2.1.).

These considerations must subsequently be supplemented by investigations on adjointness (1.2.2.) .

After that the (infinite-dimensional) analytical problem, which leads to the desired expansion theorem, can be treated in general form (1.2.3.).

The "symmetric" case $A = -B$ is of particular importance. It arises with $\xi = i\xi'$ and occurs also in other examples, for instance for the prolate spheroidal coordinates. This case leads to interesting additions, also for the algebraic problem (1.2.4.).

Finally, the application of the results to Mathieu- and spheroidal functions is at least indicated by way of examples (1.2.5.).

1.2.1. The algebraic problem.

Let $\mathcal{R}$ and $\mathcal{T}$ be linear spaces over $\mathbb{C}$ of the dimensions $k < \infty$ and $\ell < \infty$, respectively, and let

$$A : \mathcal{R} \longrightarrow \mathcal{R}, \quad B : \mathcal{T} \longrightarrow \mathcal{T}$$

be linear mappings (endomorphisms). Further let $\mathcal{J} = \mathcal{R} \otimes \mathcal{T}$ be the tensor product of these spaces. That is to say: $\mathcal{J}$ is a linear space over $\mathbb{C}$ of the dimension $k \cdot \ell$,

$$\otimes : \mathcal{R} \times \mathcal{T} \longrightarrow \mathcal{J}$$

is bilinear and

$$\mathcal{J} = \text{span}\{a \otimes b : a \in \mathcal{R}, b \in \mathcal{T}\} \quad .$$

Then the $k \cdot \ell$ $\otimes$ - "products" of k basis elements of $\mathcal{R}$ with ℓ basis elements of $\mathcal{T}$ always form a basis of $\mathcal{J}$.

Finally let

$$C : \mathcal{J} \longrightarrow \mathcal{J}$$

be the - uniquely existing - linear mapping with the property

$$C(a \otimes b) = Aa \otimes b + a \otimes Bb$$

for $a \in \mathcal{R}$, $b \in \mathcal{T}$, in short

$$C = A \otimes \text{id}_{\mathcal{T}} + \text{id}_{\mathcal{R}} \otimes B \ .$$

Such a C will be called (algebraically) simply separated.

Now there exists a basis of $\mathcal{R}$ with principal vectors of A , with respect to which A is represented by a matrix in the Jordan normal form, analogously for $\mathcal{T}$ and B . There arises the question how the normal form of C is obtained from the normal forms of A and B , and more precisely, how one has to choose a corresponding basis of $\mathcal{J}$.

This problem can immediately be largely reduced. On the one hand, the tensor product of invariant subspaces with respect to A and B is obviously again invariant with respect to C : therefore one can restrict attention to the case that A and B as normal forms have one simple Jordan box each. On the other hand, $A - \alpha\,\text{id}_{\mathcal{R}}$ and $B - \beta\,\text{id}_{\mathcal{T}}$ yield just $C - (\alpha+\beta)\text{id}_{\mathcal{J}}$: this permits to restrict oneself throughout to Jordan boxes pertaining to the eigenvalue 0 .

In this sense we start with the assumption that

$$x, Ax, \ldots, A^{k-1}x \quad \text{is basis of } \mathcal{R} \ ,$$

$$y, By, \ldots, B^{\ell-1}y \quad \text{is basis in } \mathcal{T}$$

with

$$A^k x = 0,\ B^\ell y = 0\ .$$

In addition one can assume - interchange k and ℓ if necessary -

$$\infty > k \geqq \ell \geqq 1\ .$$

We introduce the expedient abbreviations

$$x_\kappa := A^\kappa x \qquad (\kappa=0,1,2,\ldots)$$

$$y_\lambda := B^\lambda y \qquad (\lambda=0,1,2,\ldots)$$

and here with

$$(\kappa,\lambda) := x_\kappa \otimes y_\lambda\ .$$

Then one has

(1) $$C(\kappa,\lambda) = (\kappa+1,\lambda) + (\kappa,\lambda+1)\ .$$

Moreover, one will for $s \geqq 0$ construct the "homogeneous" subspaces of $\mathfrak{T}$

$$\mathfrak{H}_s := \text{span}\{(\kappa,\lambda) : \kappa + \lambda = s\}\ .$$

It is evident that

$$\dim \mathfrak{H}_s = \begin{cases} s+1 & (0 \leqq s \leqq \ell-1)\ , \\ \ell & (\ell-1 \leqq s \leqq k-1), \\ k+\ell-1-s & (k-1 \leqq s \leqq k+\ell-2). \end{cases}$$

$\mathfrak{T}$ is the direct sum of the $\mathfrak{H}_s$ $(0 \leqq s \leqq k+\ell-2)$ and with (1) there holds

(2) $$C\mathfrak{H}_s \subset \mathfrak{H}_{s+1}\ ,$$

where, of course, $\mathfrak{H}_s = \{0\}$ $(s \geqq k+\ell-1)$. More precisely, (1) yields

(3) $$C(\alpha_0(s,0) + \alpha_1(s-1,1) + \ldots + \alpha_s(0,s)) =$$
$$= \alpha_0(s+1,0) + (\alpha_0+\alpha_1)(s,1) + \ldots + (\alpha_{s-1}+\alpha_s)(1,s) + \alpha_s(0,s+1)\ .$$

Subsequently to (3) we consider real $\alpha_0,\alpha_1,\ldots,\alpha_s$ and remark that

(4) $$Cs(\alpha_0,\alpha_0+\alpha_1,\ldots,\alpha_{s-1}+\alpha_s,\alpha_s) \leqq Cs(\alpha_0,\alpha_1,\ldots,\alpha_s)\ ,$$

where $Cs(\quad)$ is the number of changes of sign in the real sequence enclosend within the parentheses. The simple proof is postponed to the end of this section.

Now one determines, suitably by meancs of (3), the homogeneous eigenvectors of C to 0. It is recognized that here, apart from a constant factor, only the ℓ following are possible:

(5)
$$e_0 := (k-1,\ell-1) \in \mathfrak{H}_{k+\ell-2}\ ,$$
$$e_1 := (k-1,\ell-2) - (k-2,\ell-1) \in \mathfrak{H}_{k+\ell-3}\ ,$$
$$e_2 := (k-1,\ell-3) - (k-2,\ell-2) + (k-3,\ell-1) \in \mathfrak{H}_{k+\ell-4}\ ,$$
$$\vdots$$
$$e_{\ell-1} := (k-1,0) - + \ldots + (-1)^{\ell-1}(k-\ell,\ell-1) \in \mathfrak{H}_{k-1}\ .$$

Now it is easy to obtain the normal form of C. One starts with

$$z_o := \binom{k+\ell-2}{k-1}^{-1} \quad (0,0)$$

and finds

$$C^{k+\ell-2} z_o = e_o .$$

If $\ell \geqq 2$ then

$$C^{k+\ell-3} \mathfrak{H}_1 \subset \mathfrak{H}_{k+\ell-2} ,$$

$$\dim \mathfrak{H}_1 = 2 , \dim \mathfrak{H}_{k+\ell-2} = 1 .$$

Therefore there exists

$$0 \neq z_1 \in \mathfrak{H}_1 , C^{k+\ell-3} z_1 = 0 .$$

If ν is maximal with $C^{\nu} z_1 \neq 0$, then $C^{\nu} z_1 \in \mathfrak{H}_{\nu+1}$ and is an eigenvector. It must be equal to e_1 apart from a factor. For z_1 can be chosen with real coefficients and has one change of sign at the most. Repeated application of (4) yields the same result for $C^{\nu} z_1$. But the representation of e_ρ has exactly ρ changes of sign; this proves the assertion. Therefore we can determine

$$z_1 \in \mathfrak{H}_1 , C^{k+\ell-4} z_1 = e_1 .$$

We outline a further step for $\ell \geqq 3$. Because of

$$C^{k+\ell-5} \mathfrak{H}_2 \subset \mathfrak{H}_{k+\ell-3} ,$$

$$\dim \mathfrak{H}_2 = 3 , \dim \mathfrak{H}_{k+\ell-3} = 2 ,$$

there exists

$$0 \neq z_2 \in \mathfrak{H}_2 , C^{k+\ell-5} z_2 = 0 .$$

If ν is maximal with $C^{\nu} z_2 \neq 0$, so is $C^{\nu} z_2 \in \mathfrak{H}_{\nu+2}$ and is an eigenvector. Now z_2, if chosen with real coefficients, has only two changes of sign at the most; therefore repeated application of (4) and consideration of (5) yields

$$z_2 \in \mathfrak{H}_2 , C^{k+\ell-6} z_2 = e_2 .$$

One continues in this manner with the determination of

$$z_s \in \mathfrak{H}_s , C^{k+\ell-2s-2} z_s = e_s \qquad (s=0,1,2,\dots,\ell-1).$$

Thus the following result is obtained:

(i) For $s = 0,1,\dots,\ell-1$

$$C^s z_o , C^{s-1} z_1, \dots, C z_{s-1}, z_s$$

form a basis of $\mathfrak{H}_s$.

(ii) For $s = \ell - 1,\dots,k - 1$

$$C^s z_o , C^{s-1} z_1, \dots, C^{s-\ell+1} z_{\ell-1}$$

form a basis of $\mathfrak{H}_s$,

(iii) For $s = k - 1, \ldots, k + \ell - 2$

$$C^s z_0, \; C^{s-1} z_1, \ldots, C^{2s-k-\ell+2} z_{k+\ell-2-s} = e_{k+\ell-2-s}$$

form a basis of $\mathfrak{H}_s$.

(iv) For $0 \leqq 2s \leqq k + \ell - 2$

$$C^{k+\ell-2s-2} : \mathfrak{H}_s \longrightarrow \mathfrak{H}_{k+\ell-2-s}$$

is bijective.

(v) The normal form of C consists of ℓ Jordan boxes whose numbers of rows are $k + \ell - 1, k + \ell - 3, \ldots, k - \ell + 1$. It is obtained from the principal vector z_0 of order $k + \ell - 1$ to the eigenvector e_0 , from the principal vector z_1 of order $k + \ell - 3$ to the eigenvector e_1 etc..

We state these facts appropriately in the following scheme:

$$\begin{array}{ccccccccccccccc}
\mathfrak{H}_o & \mathfrak{H}_1 & \mathfrak{H}_2 & \cdots & \mathfrak{H}_{\ell-1} & \cdots & \mathfrak{H}_{k-1} & \cdots & \mathfrak{H}_{k+\ell-4} & \cdots & \mathfrak{H}_{k+\ell-3} & \mathfrak{H}_{k+\ell-2} \\
z_0 & Cz_0 & C^2 z_0 & \cdots & C^{\ell-1} z_0 & \cdots & C^{k-1} z_0 & \cdots & C^{k+\ell-4} z_0 & \cdots & C^{k+\ell-3} z_0 & e_o \\
 & z_1 & Cz_1 & \cdots & C^{\ell-2} z_1 & \cdots & C^{k-2} z_1 & \cdots & C^{k+\ell-5} z_1 & \cdots & e_1 & \\
 & & z_2 & \cdots & C^{\ell-3} z_2 & \cdots & C^{k-3} z_2 & \cdots & e_2 & & & \\
 & & & \ddots & \vdots & & \vdots & \cdot^{\cdot^{\cdot}} & & & & \\
 & & & & z_{\ell-1} & \cdots & e_{\ell-1} & . & & & &
\end{array}$$

Now we return to the general case of arbitrary endomorphisms A and B of finite-dimensional spaces $\mathfrak{R}$ and $\mathfrak{T}$, respectively, over $\mathbb{C}$.

On the strength of the remark on reduction made above, there holds:

To each pair of a Jordan box for A to the eigenvalue α and of a Jordan box for B to the eigenvalue β with the numbers of rows κ and λ , respectively, there correspond $\min(\kappa,\lambda)$ Jordan boxes for C to the eigenvalue $\alpha + \beta$ with the numbers of rows $\kappa + \lambda - 1, \kappa + \lambda - 3, \ldots, |\kappa - \lambda| + 1$. The collection of all these yields the normal form of C .

Our presentation furnishes the instruction to the practical determination of the appertaining principal vectors following the rules of linear algebra. It is further remarked that the statement (iv) is equivalent to the non-vanishing of certain determinants, which are built up from binomial coefficients. Incidentally, in the case $k = \ell$ they are found to be Gram determinants.

The remains to append the proof of (4).
Obviously, (4) is a special case of

$$\text{(6)} \qquad Cs(\alpha_1 + \alpha_2, \alpha_2 + \alpha_3, \ldots, \alpha_n + \alpha_{n+1}) \leqq Cs(\alpha_1, \alpha_2, \ldots, \alpha_{n+1})$$

with $\alpha_1 = \alpha_{n+1} = 0$ and with a corresponding change of notation. We shall give a proof of (6) . This is certainly correct for $n = 1$ and $n = 2$, and on the other hand for arbitrary n, if the number of changes of sign in the right member is $n-1$ or n. But (6) is also correct if the number of changes of sign in the right member is $n-2$ while $Cs(\alpha_1,\alpha_2) = Cs(\alpha_n,\alpha_{n+1}) = 0$ and $\alpha_1 + \alpha_2$ has the sign of $\alpha_2 \neq 0$ and $\alpha_n + \alpha_{n+1}$ has the sign of $\alpha_n \neq 0$; then $n-2$ changes of sign in the right member cannot induce $n-1$ changes of sign in the left member. In all other cases there is a pair (α_k,α_{k+1}), $1 < k < n$, without a change of sign. But then for $\alpha_k + \alpha_{k+1} \neq 0$

$$Cs(\alpha_1,\alpha_2,\ldots,\alpha_{n+1}) = Cs(\alpha_1,\ldots,\alpha_{k+1}) + Cs(\alpha_k,\ldots,\alpha_{n+1})$$

and by induction one concludes to

$$\geqq Cs(\alpha_1 + \alpha_2,\ldots,\alpha_k + \alpha_{k+1}) + Cs(\alpha_k + \alpha_{\ell+1},\ldots,\alpha_n + \alpha_{n+1})$$

$$= Cs(\alpha_1 + \alpha_2,\ldots,\alpha_n + \alpha_{n+1}) ,$$

and for $\alpha_k = \alpha_{k+1} = 0$ one has evidently a direct reduction $n+1 \longrightarrow n$.

1.2.2. Adjoint mappings.

We start directly from 1.2.1.. In addition we assume: $\mathcal{R}^*$ and $\mathcal{T}^*$ are linear spaces over $\mathbb{C}$ with

$$\dim \mathcal{R}^* = \dim \mathcal{R} , \ \dim \mathcal{T}^* = \dim \mathcal{T} .$$

Let non-degenerate bilinear mappings of $\mathcal{R} \times \mathcal{R}^*$ into $\mathbb{C}$ and of $\mathcal{T} \times \mathcal{T}^*$ into $\mathbb{C}$ be given which in both cases are denoted by $\langle\ ,\ \rangle$. It follows from the non-degeneracy requirement that

$$\langle \cdot , r^* \rangle \longleftrightarrow r^* \in \mathcal{R}^*$$

and

$$\langle \cdot , s^* \rangle \longleftrightarrow s^* \in \mathcal{T}^* ,$$

respectively, represent isomorphisms to the dual space. Finally let A^* and B^* be the adjoint endomorphisms of $\mathcal{R}^*$ and $\mathcal{T}^*$, respectively, in regard of the pertinent $\langle\ ,\ \rangle$:

$$\langle Au,v^* \rangle = \langle u,A^*v^* \rangle \qquad (u \in \mathcal{R},\ v^* \in \mathcal{R}^*) ,$$

$$\langle Bu,v^* \rangle = \langle u,B^*v^* \rangle \qquad (u \in \mathcal{T},\ v^* \in \mathcal{T}^*) .$$

If now one forms, as in 1.2.1., the tensor product $\mathcal{J}^* = \mathcal{R}^* \otimes \mathcal{T}^*$, then this can be considered as the dual space to $\mathcal{J} = \mathcal{R} \otimes \mathcal{T}$: that is to say, there exists obviously in a unique way a bilinear mapping, again denoted by $\langle\ ,\ \rangle$, from $\mathcal{J} \times \mathcal{J}^*$ into $\mathbb{C}$ with

(1) $$\langle a \otimes b,\ a^* \otimes b^* \rangle = \langle a,a^* \rangle \cdot \langle b,b^* \rangle$$

$$(a \in \mathcal{R},\ b \in \mathcal{T},\ a^* \in \mathcal{R}^*,\ b^* \in \mathcal{T}^*) .$$

If then one defines, as in 1.2.1.,

(2) $$C^* := A^* \otimes \mathrm{id}_{\mathcal{T}^*} + \mathrm{id}_{\mathcal{R}^*} \otimes B^* ,$$

then C^* is adjoint to C with respect to the $\langle\ ,\ \rangle$ which had just been introduced:

(3) $$\langle Cc,c^*\rangle = \langle c,C^*c^*\rangle \qquad (c \in \mathcal{T},\ c^* \in \mathcal{T}^*) .$$

This has to be verified only for $c = a \otimes b,\ c^* = a^* \otimes b^*$:

$$\langle Cc,c^*\rangle = \langle Aa \otimes b + a \otimes Bb,\ a^* \otimes b^*\rangle =$$

$$= \langle Aa,a^*\rangle\langle b,b^*\rangle + \langle a,a^*\rangle\langle Bb,b^*\rangle = \langle a,A^*a^*\rangle\langle b,b^*\rangle + \langle a,a^*\rangle\langle b,B^*b^*\rangle =$$

$$= \langle a \otimes b,\ A^*a^* \otimes b^* + a^* \otimes B^*b^*\rangle = \langle c,C^*c^*\rangle \quad .$$

Now let $x_o,x_1,\ldots,x_{k-1}$ be a basis of $\mathcal{R}$ with respect to which A is represented by a Jordan normal form with boxes

$$\begin{pmatrix} \alpha & & & 0 \\ 1 & \ddots & & \\ & \ddots & \ddots & \\ 0 & & 1 & \alpha \end{pmatrix} .$$

Then there exists to it uniquely a dual basis $x_o^*,x_1^*,\ldots,x_{k-1}^*$ of $\mathcal{R}^*$:

$$\langle x_\nu,x_\mu^*\rangle = \delta_{\nu\mu} \qquad (\nu,\mu = 0,1,\ldots,k-1) .$$

In this basis A^* is represented by the corresponding transposed matrix, that is with boxes

$$\begin{pmatrix} \alpha & 1 & & 0 \\ & \ddots & \ddots & \\ & & \ddots & 1 \\ 0 & & & \alpha \end{pmatrix} .$$

Thus one obtains a biorthogonal system of corresponding principal vectors of A and A^* which serve as bases for $\mathcal{R}$ and $\mathcal{R}^*$, respectively.

Analogous results hold for $\mathcal{T},\mathcal{T}^*,B,B^*$.

In the following, it will be shown how the procedure for the choice of a basis for the normal form of C in <u>1.2.1.</u>, and in analogy for C^*, almost immediately furnishes a suitable biorthogonal system. For that purpose we start from the just described dual bases for A,A^* and B,B^*, respectively, and for the detailed consideration we restrict ourselves to a pair of single Jordan boxes and, without loss of generality, to the eigenvalues 0 and with $k \geqq \ell$.

Then we have the basis

$$x,Ax,\ldots,A^{k-1}x$$

of $\mathcal{R}$ with $A^kx = 0$, and also the dual basis

$$A^{*k-1}x^*,\ldots,A^*x^*,x^*$$

of $\mathcal{R}^*$ with $A^{*k}x^* = 0$ and analogously the basis

$$y,By,\ldots,B^{\ell-1}y$$

of $\mathcal{T}$ with $B y = 0$, and also the dual basis

$$B^{*\ell-1}y^*,\ldots,B^*y^*,y^*$$

of $\mathcal{T}^*$ with $B^{*\ell}y^* = 0$. With the earlier notation

$$x_\kappa = A^\kappa x, \qquad y_\kappa = B^\kappa y ,$$
$$x^*_\lambda = A^{*\lambda}x^*, \qquad y^*_\lambda = B^{*\lambda}y^* ,$$

one has

$$\langle x_i, x^*_{k-1-j}\rangle = \delta_{ij} , \langle y_r, y^*_{\ell-1-s}\rangle = \delta_{ij}$$

and consequently with (1)

(4) $$\langle x_i \otimes y_r, x^*_{k-1-j} \otimes y^*_{\ell-1-s}\rangle = \delta_{ij} \cdot \delta_{rs} .$$

Therefore, we have already here dual bases for $\mathcal{T}, \mathcal{T}^*$. Now we evaluate according to 1.2.1. the eigenvectors $e_0, e_1, \ldots, e_{\ell-1}$ and along with them the principal vectors $z_0, z_1, \ldots, z_{\ell-1}$ for C, and analogously, $e^*_0, e^*_1, \ldots, e^*_{\ell-1}$, $z^*_0, z^*_1, \ldots, z^*_{\ell-1}$ for C^*.
Then we assert:

Theorem: If one writes for C^* the scheme in 1.2.1. in the reversed order of columns and provides the rows with the factors

$$\zeta_\rho := \langle z_\rho, e^*_\rho\rangle^{-1} \qquad (\rho = 0,1,\ldots,\ell-1) :$$

$$\begin{array}{ccccccc}
\mathcal{B}^*_{k+\ell-2} & \mathcal{B}^*_{k+\ell-3} & \cdot & \ldots & \cdot & \mathcal{B}^*_1 & \mathcal{B}^*_0 \\
\zeta_0 e^*_0 & \zeta_0 C^{*k+\ell-3} z^*_0 & \cdot & \ldots & \cdot & \zeta_0 C^* z^*_0 & \zeta_0 z^*_0 \\
 & \zeta_1 e^*_1 & \cdot & \ldots & \cdot & \zeta_1 z^*_1 & \\
 & & \zeta_2 e^*_2 & \ldots & \zeta_2 z^*_2 & & \\
 & & & \vdots & & &
\end{array} ,$$

then one obtains the dual basis to the scheme for C in 1.2.1.: for equally positoned terms we have $\langle\cdot,\cdot\rangle = 1$, else $\langle\cdot,\cdot\rangle = 0$. -

For the proof we remark at first that

$$\langle z_\rho, e^*_\rho\rangle \neq 0 .$$

This follows at once because of the alternating signs in the representations of the z_ρ and of the e^*_ρ, respectively, according to the remarks in 1.2.1. ; however, this can also be inferred from the orthogonality which is in force apart from that. Moreover, it is immediately established according to (4) that

$$\langle \mathcal{B}_s, \mathcal{B}^*_{k+\ell-2-r}\rangle = 0 \qquad (r \neq s) .$$

Therefore one only needs to consider pairs of elements from $\mathcal{B}_s$ and $\mathcal{B}^*_{k+\ell-2-s}$, that is from the same column of both schemes. Here one starts, for instance, from $C^{*k+\ell-3-s} z^*_1 = 0$ and concludes therefrom with (3) to

$$\langle C^s z_0, C^{*k+\ell-3-s} z^*_1\rangle = 0$$

for $1 \leq s \leq k + \ell - 3$; analogously, one obtains

$$\langle C^s z_1, C^{*k+\ell-3-s} z^*_o\rangle = 0 .$$

Correspondingly, it obviously follows from

$$C^{*k+\ell-5}z_2^* = 0$$

that

$$\langle C^{s+1}z_0, C^{*k+\ell-5-s}z_2^*\rangle = 0$$

and

$$\langle C^s z_1, C^{*k+\ell-5-s}z_2^*\rangle = 0$$

for $1 \leq s \leq k + \ell - 5$ and conversely. Continuing to conclude in this way one obtains all the assertions $\langle\cdot,\cdot\rangle = 0$.

There remains the verification of

$$(0\neq)\ \langle C^s z_\rho, C^{*k+\ell-2-s-2\rho}z_\rho^*\rangle = \langle z_\rho, e_\rho^*\rangle = \zeta_\rho^{-1} ,$$

which is again obtained with (3) for $0 \leq s \leq k + \ell - 2\rho - 2$.

Proceeding in this manner for every pair of "boxes" for A and B, respectively, and for A^* and B^*, respectively, one obtains, in summarizing, dual bases of $\mathcal{T}$ and $\mathcal{T}^*$ for the normal form of C and the transposed normal form of C^* .

1.2.3. The analytical problem. Expansion theorem.

The analytical problems, which were described in the introduction 1.2.0., shall now be treated in an abstract theory on the basis of 1.2.1., 1.2.2., which includes the finite dimensional case and lets emerge the underlying structures. We start with the following assumptions (1) to (5) .

(1) Let $\mathcal{R}, \mathcal{R}^*, \mathcal{S}, \mathcal{S}^*, \mathcal{T}, \mathcal{T}^*$ be linear spaces over $\mathbb{C}$, all of them $\neq \{0\}$; let

$$\langle\ ,\ \rangle : \mathcal{R} \times \mathcal{R}^* \longrightarrow \mathbb{C}$$
$$\langle\ ,\ \rangle : \mathcal{S} \times \mathcal{S}^* \longrightarrow \mathbb{C}$$
$$\langle\ ,\ \rangle : \mathcal{T} \times \mathcal{T}^* \longrightarrow \mathbb{C}$$

be bilinear mappings which are non-degenerate in the following sense: if for all $u \in \mathcal{R}$ $\langle u,v^*\rangle = 0$, then $v^* = 0$ and conversely; analogously in the other cases.

Let

$$\otimes : \mathcal{R} \times \mathcal{S} \longrightarrow \mathcal{T} \quad ;$$
$$\otimes : \mathcal{R}^* \times \mathcal{S}^* \longrightarrow \mathcal{T}^*$$

be bilinear mappings which are non-degenerate in the following sense: if $x_1, x_2, \ldots, x_r$ are linearly independent in $\mathcal{R}$ and $\mathcal{R}^*$, respectively, and if $y_1, y_2, \ldots, y_s$ are linearly independent in $\mathcal{S}$ and $\mathcal{S}^*$, respectively, then the $r \cdot s$ elements $x_\rho \otimes y_\sigma$ are linearly independent in $\mathcal{T}$ and $\mathcal{T}^*$, respectively; for $a \in \mathcal{R}$, $b \in \mathcal{S}$, $c^* \in \mathcal{R}^*$, $d^* \in \mathcal{S}^*$ there shall always hold

$$\langle a \otimes b, c^* \otimes d^*\rangle = \langle a,c^*\rangle \cdot \langle b,d^*\rangle .$$

(2) Let $\mathcal{R}$ and $\mathcal{T}$ be seperated-topological linear spaces; let

$$\langle\ ,\ \rangle_1 : \mathcal{T} \times \mathcal{S}^* \longrightarrow \mathcal{R}$$

be a bilinear mapping, continuous in the first variable, with the property that for $a \in \mathcal{R}$, $b \in \mathcal{T}$, $d^* \in \mathcal{T}^*$ then holds

$$\langle a \otimes b, d^* \rangle_1 = \langle b, d^* \rangle \, a \; .$$

$\langle \; , \; \rangle : \mathcal{Z} \times \mathcal{Z}^* \to \mathbb{C}$ as well as $\otimes : \mathcal{R} \times \mathcal{T} \to \mathcal{Z}$ shall be continuous in the first variable.

(3) Let

$$A : \mathcal{R} \longrightarrow \mathcal{R} , \qquad B : \mathcal{T} \longrightarrow \mathcal{T} ,$$

$$A^* : \mathcal{R}^* \longrightarrow \mathcal{R}^* , \qquad B^* : \mathcal{T}^* \longrightarrow \mathcal{T}^*$$

be endomorphisms ; let A, A^* and B, B^*, respectively, be adjoint with regard to the related $\langle \; , \; \rangle$, and let A be continuous.

(4) Let

$$A \otimes \mathrm{id}_{\mathcal{T}} : \mathcal{Z} \longrightarrow \mathcal{Z} ,$$

$$\mathrm{id}_{\mathcal{R}} \otimes B \; : \mathcal{Z} \longrightarrow \mathcal{Z}$$

be continuous endomorphisms with

$$(A \otimes \mathrm{id}_{\mathcal{T}}) \; (a \otimes b) = Aa \otimes b \; ,$$

$$(\mathrm{id}_{\mathcal{R}} \otimes B) \; (a \otimes b) = a \otimes Bb$$

for $a \in \mathcal{R}$, $b \in \mathcal{T}$. Also let

$$C := A \otimes \mathrm{id}_{\mathcal{T}} + \mathrm{id}_{\mathcal{R}} \otimes B$$

and

$$C^* : \mathcal{Z}^* \longrightarrow \mathcal{Z}^*$$

be an endomorphism which is adjoint to C with respect to $\langle \; , \; \rangle$.

(5) For every eigenvalue of A the associated principal spaces of A and A^* have the same finite dimension; analogously for B and B^*. B has at most a countable number of eigenvalues. For every eigenvalue β of B

$$y_\nu(\beta), \; y^*_\mu(\beta) \qquad (\nu, \mu = 1, 2, \ldots, o(\beta))$$

shall be dual bases of the principal spaces of B and B^*, respectively, to β. With a fixed succession of the eigenvalues β to B, there shall hold the expansion

$$c = \sum_{\beta} \sum_{n=1}^{o(\beta)} \langle c, y^*_n(\beta) \rangle_1 \otimes y_n(\beta)$$

for every $c \in \mathcal{Z}$. The linear combinations of the principal vectors of A shall be dense in $\mathcal{R}$.

We start with the

<u>Lemma 1 :</u> For $c^* \in \mathcal{R}^*$, $d^* \in \mathcal{T}^*$ there holds

$$C^*(c^* \otimes d^*) = A^* c^* \otimes d^* + c^* \otimes B^* d^* \; .$$

Proof: Let δ^* be the difference of the two members. Then the last requirement of ① yields at once

$$\langle c,\delta^*\rangle = 0$$

for $c = a \otimes b$, $a \in \mathcal{R}$, $b \in \mathcal{T}$, and consequently for all finite linear combinations of such c. But according to ② $\langle\ ,\ \rangle$ is continuous in the first variable, and according to ⑤ the linear combinations so obtained are dense in $\mathcal{Z}$; consequently $\langle c,\delta^*\rangle = 0$ for all $c \in \mathcal{Z}$. Because of the non-degeneracy requirement in ① this yields $\delta^* = 0$.

Further we formulate

Lemma 2 : For $w \in \mathcal{Z}$, $y^* \in \mathcal{T}^*$ there holds

$$A\langle w,y^*\rangle_1 = \langle (A \otimes id_{\mathcal{T}})w,y^*\rangle_1 .$$

Proof : From ② and ④ this is verified for $w = a \otimes b$, $a \in \mathcal{R}$, $b \in \mathcal{T}$. Because of linearity, the assertion holds for all finite linear combinations of such $a \otimes b$. Now these are dense in $\mathcal{Z}$ as assumed in ⑤, $\langle\ ,\ \rangle_1$ is continuous in the first variable as assumed in ③, A and $(A \otimes id_{\mathcal{T}})$ are continuous according to ③ and ④, respectively. Thus the assertion holds true for all $w \in \mathcal{Z}$.

Almost in the same manner one obtains

Lemma 3 : For $w \in \mathcal{Z}$, $y^* \in \mathcal{T}$ there holds

$$\langle (id_{\mathcal{R}} \otimes B)w,y^*\rangle_1 = \langle w,B^*y^*\rangle_1 .$$

Here one has also to use that B and B^* are adjoint with respect to $\langle\ ,\ \rangle$.

Of essential importance is now

Theorem 4 : For $w \in \mathcal{Z}$, $y^* \in \mathcal{T}^*$ there holds

$$A\langle w,y^*\rangle_1 = \langle Cw,y^*\rangle_1 - \langle w,B^*y^*\rangle_1$$

Let $n \geqq 0$, $m \geqq 0$, $m + n \geqq 1$; then $(C-(\alpha+\beta))^m w = 0$ and $(B^*-\beta)^n y^* = 0$ always yield $(A-\alpha)^{n+m-1}\langle w,y^*\rangle_1 = 0$.

Proof : The first statement follows directly from Lemma 2 and Lemma 3. For the second statement it suffices to consider the case $\alpha = \beta = 0$. It is obviously correct for $m + n = 1$; for $n + m \geqq 2$ it is proved by induction with respect to $n + m$: this gives already

$$A^{m+n-2}\langle Cw,y^*\rangle_1 = A^{m+n-2}\langle w,B^*y^*\rangle_1 = 0 .$$

Application of A^{m+n-2} to the first statement yields the assertion.

Elementary algebra yields the

Lemma 5: If y is principal vector of B to the eigenvalue β and y^* is principal vector of B^* to the eigenvalue $\beta^* \neq \beta$, then $\langle y,y^*\rangle = 0$. Analogous results hold for A,A^* and B,B^*.

Proof : Let $(B - \beta)^n y = 0$ and $(B^* - \beta^*)^m y^* = 0$. The polynomials $(x - \beta)^n$ and $(x - \beta^*)^m$ have no common factor. Therefore there exist complex polynomials $p(x)$, $q(x)$ with

$$1 = p(x)(x-\beta)^n + (x-\beta^*)^m q(x) .$$

If x is replaced by B, then

$$1 = p(B)(B-\beta)^n + (B-\beta^*)^m q(B) ;$$

from which we deduce

$$\langle y,y^*\rangle = \langle p(B)(B-\beta)^n y,y^*\rangle + \langle q(B)y,(B^*-\beta^*)^m y^*\rangle = 0$$

We also note

Lemma 6 : If α is eigenvalue to A, then to each principal vector $x^* \neq 0$ of A^* to α there exists a principal vector x of A to α with $\langle x,x^*\rangle \neq 0$, and vice versa.

Lemma 7 : The linear combinations of elements $x \otimes y$, with x principal vector to A and y principal vector to B, which are consequently principal vectors of C, are dense in $\mathfrak{Z}$.

Proof : Lemma 7 follows from the last two requirements of (5), from the continuity of occurring $\otimes$ in the first variable ((2)), and from the continuity of $+$ in topological linear spaces. - For the proof of Lemma 6 one can now use Lemma 7 and the continuity of $\langle\ ,\ \rangle : \mathfrak{Z} \times \mathfrak{Z}^* \longrightarrow \mathbb{C}$ according to (2), and also draw upon Lemma 5 : Let $x^* \neq 0$ be principal vector of A^* to α and $y^* \neq 0$ be principal vector to an eigenvalue β of B^*, which exists because of $\mathfrak{Z} \neq \{0\}$. If then x^* were orthogonal to all principal vectors of A to α, then one could conclude with Lemma 5 and with the last line of (1) that $x^* \otimes y^*$ were orthogonal to all those principal vectors of C which were mentioned in Lemma 7. This Lemma and the continuity of $\langle\ ,\ \rangle$ according to (2) would then give

$$\langle c,x^* \otimes y^*\rangle = 0 \qquad (c \in \mathfrak{Z})$$

in contradiction to the non-degeneracy requirement for $\langle\ ,\ \rangle : \mathfrak{Z} \times \mathfrak{Z}^* \longrightarrow \mathbb{C}$. - The inverted statement follows in known manner from the fact that the dimensions of both principal spaces are finite and equal according to (5).

The main result of our considerations is the following theorem on the series representation of all principal vectors of C :

Theorem 8 : Let w be principal vector of C to the eigenvalue γ.

(i) Then there exists a series representation

$$(*) \qquad w = \sum_{\beta} w(\beta) ,$$

$$(**) \qquad w(\beta) = \sum_{\nu=1}^{k(\beta)\ell(\beta)} \langle w,w_\nu^*(\beta)\rangle w_\nu(\beta) .$$

Here the sum $(*)$ is to be extended over those β in the sequence of (5), to which $\gamma - \beta$ is eigenvalue of A.

If $\ell(\beta)$ is the dimension of the principal space of B to β and $k(\beta)$ is the dimension of the principal space of A to $\gamma - \beta$, then

$$w_1(\beta),\dots,w_{k(\beta)\cdot\ell(\beta)}(\beta)$$

shall be a basis of the tensor product of the mentioned principal spaces according to 1.2.1. and

$$w_1^*(\beta),\dots,w^*_{k(\beta)\ell(\beta)}(\beta)$$

shall be the associated dual basis of the tensor product of the corresponding principal spaces of A^* and B^* according to 1.2.2..

(ii) If, in particular, $(C-\gamma)^m w = 0$, $m \in \mathbb{N}$, then in (**) only those $w_\nu(\beta)$ occur with non-vanishing coefficients for which also $(C-\gamma)^m w_\nu(\beta) = 0$.

(iii) If, conversely, $m \in \mathbb{N}$, $\gamma \in \mathbb{C}$, with $\eta_\nu(\beta) \in \mathbb{C}$

$$w = \sum_\beta w(\beta) ,$$

$$w(\beta) = \sum_{\nu=1}^{k(\beta)\ell(\beta)} \eta_\nu(\beta) w_\nu(\beta) ,$$

then there always holds

$$\eta_\nu(\beta) = \langle w, w_\nu^*(\beta)\rangle .$$

If, in particular, $\eta_\nu(\beta) = 0$ in case $(C-\gamma)^m w_\nu(\beta) \neq 0$, then there is also $(C-\gamma)^m w = 0$.

Proof : If one considers the expansion of w according to (5), then due to Theorem 4 the $\langle w, y_n^*(\beta)\rangle_1$ are principal vectors of A to the eigenvalue $\gamma - \beta$. Now one can apply the considerations of 1.2.1., 1.2.2. to the principal spaces of A to $\gamma - \beta$, of B to β, and to the corresponding ones of A^* and B^* : take notice of (5), (1), Lemma 6, (4), Lemma 1. Due to that

$$w(\beta) = \sum_{n=0}^{o(\beta)} \langle w, y_n^*(\beta)\rangle_1 \otimes y_n(\beta)$$

can be represented in the form

$$w(\beta) = \sum_{\nu=1}^{k(\beta)\ell(\beta)} \eta_\nu(\beta) w_\nu(\beta) .$$

Now, due to (2), $\langle\ ,\ \rangle : \mathcal{T} \times \mathcal{T}^* \longrightarrow \mathbb{C}$ is continuous in the first variable; therefore Lemma 5 (for C and C^*) gives just the representation (**). For (ii) one has just to look a little more precisely at the schemes in 1.2.1. and 1.2.2.. One recognizes that the dual elements to the basis elements chosen there, which do not vanish upon application of C^m, lie just in the range of C^{*m}. But then one has

$$\langle w, C^{*m}u^*\rangle = \langle C^m w, u^*\rangle = 0 .$$

Finally, (iii) follows immediately from the continuity of $\langle \, , \, \rangle$ and C. - We remark only that for (ii) also Theorem 4 could be utilized in more detail.

1.2.4. The symmetric case.

In the following we consider the case

$$\mathcal{R} = \mathcal{T}\,, \quad \mathcal{R}^* = \mathcal{T}^* \;,$$
$$B = -A, \quad B^* = -A^* \;,$$

that is to say

$$C = A \otimes id_{\mathcal{R}} - id_{\mathcal{R}} \otimes A, \quad C^* = A^* \otimes id_{\mathcal{R}^*} - id_{\mathcal{R}^*} \otimes A^* \,.$$

In addition to ① - ⑤ in 1.2.3. we assume

⑥ Let $S : \mathcal{T} \longrightarrow \mathcal{T}$ be continuous and linear with

$$S(a \otimes b) = b \otimes a$$

for $a, b \in \mathcal{R}$.

Then we decompose directly

$$\mathcal{T} = \mathcal{T}^+ \dotplus \mathcal{T}^-$$

into the subspaces of the symmetric elements $\mathcal{T}^+$, with $Sw = w$, and of the antisymmetric elements $\mathcal{T}^-$, with $Sw = -w$. Obviously, the associated projections are

$$P := \frac{1}{2}(id_{\mathcal{T}} + S), \quad Q := \frac{1}{2}(id_{\mathcal{T}} - S) \;,$$
$$P + Q = id_{\mathcal{T}} \;, \quad PQ = QP = 0 \;.$$

The following properties are of importance:

$$C\mathcal{T}^+ \subset \mathcal{T}^- \;, \quad C\mathcal{T}^- \subset \mathcal{T}^+ \;,$$
$$CS = -SC \;,$$
$$C^2\mathcal{T}^+ \subset \mathcal{T}^+ \;, \quad C^2\mathcal{T}^- \subset \mathcal{T}^- \qquad (1)$$
$$C^2 S = S C^2 \;.$$

We shall be interested in the solutions of $Cw = 0$ and of $C^2 w = 0$ in $\mathcal{T}^+$ and $\mathcal{T}^-$. For that one has, due to 1.2.3., Theorem 8, to consider the spaces

$$\mathcal{T}_\alpha = \mathfrak{H}_\alpha \otimes \mathfrak{H}_\alpha$$

with the principal spaces $\mathfrak{H}_\alpha$ of A to the eigenvalues α . It is expedient to decompose as above

$$\mathcal{T}_\alpha = \mathcal{T}_\alpha^+ \dotplus \mathcal{T}_\alpha^-$$

and to considerate first a corresponding splitting of the basis of $\mathcal{T}_\alpha$ which is obtained according to 1.2.1..

Let α be fixed at the moment; we use the notation

$$k = k_1 + \ldots + k_v \;,$$

where $k = \dim \mathfrak{H}_\alpha$ is the order of α , v is the number of boxes or the multiplicity of α , respectively, and $k_1,\dots,k_v$ are the numbers of rows of the various boxes or the dimensions of the associated subspaces $\mathfrak{h}_{\alpha,\nu}$ $(\nu=1,2,\dots,v)$, respectively. If, for $1 \leq \nu < \mu \leq v$, one now constructs the basis

$$w_{(\nu,\mu);\kappa} \qquad (\kappa=1,2,\dots,k_\nu \cdot k_\mu)$$

for $\mathfrak{h}_{\alpha,\nu} \otimes \mathfrak{h}_{\alpha,\mu}$ according to 1.2.1., then one obtains with

(*)
$$Pw_{(\nu,\mu);\kappa} \, , \quad Qw_{(\nu,\mu);\kappa}$$

exactly $k_\nu \cdot k_\mu$ elements of $\mathfrak{T}^+_\alpha$ and $\mathfrak{T}^-_\alpha$, respectively. If, on the other hand, one now constructs for $1 \leq \nu \leq v$ the basis

(**)
$$w_{(\nu,\nu);\kappa} \qquad (\kappa=1,2,\dots,k_\nu^2)$$

for $\mathfrak{h}_{\alpha,\nu} \otimes \mathfrak{h}_{\alpha,\nu}$ according to 1.2.1., then it is immediately recognized that these elements themselves are in $\mathfrak{T}^+_\alpha$ or in $\mathfrak{T}^-_\alpha$, according to the scheme

(***)
$$\begin{array}{ccccccccccc}
+ & - & + & & & \cdots & & & + & - & + \\
 & + & - & + & & \cdots & & & - & + & \\
 & & \ddots & & & \vdots & & & \cdot^{\cdot^{\cdot}} & & \\
 & & & & + & - & + & & & & \\
 & & & & & + & & & & &
\end{array}$$

Therefore one has here $\frac{1}{2}\,k_\nu(k_\nu+1)$ elements for $\mathfrak{T}^+_\alpha$ and $\frac{1}{2}\,k_\nu(k_\nu-1)$ elements for $\mathfrak{T}^-_\alpha$. Altogether one obtains in this manner bases for $\mathfrak{T}^+_\alpha$ and $\mathfrak{T}^-_\alpha$ with the dimension numbers being

$$\dim \mathfrak{T}^+_\alpha = \sum_{\nu<\mu} k_\nu \cdot k_\mu + \sum_\nu \frac{1}{2}\,k_\nu(k_\nu+1) = \frac{1}{2}\,k(k+1) \; ,$$

$$\dim \mathfrak{T}^-_\alpha = \sum_{\nu<\mu} k_\nu \cdot k_\mu + \sum_\nu \frac{1}{2}\,k_\nu(k_\nu-1) = \frac{1}{2}\,k(k-1) \; .$$

These considerations together with 1.2.3., Theorem 8, in which due to (6)
S,P,Q can be applied termwise, lead to the following theorems.

Theorem 1 : There exists a $0 \neq w \in \mathfrak{T}$ with $C^2w = 0$, $Cw \neq 0$ if and only if for at least one eigenvalue α of A the order is greater than the multiplicity.

Theorem 2 : The equation

$$Cw = 0 \qquad (0 \neq w \in \mathfrak{T}^-)$$

is solvable if and only if at least one eigenvalue α of A has a multiplicity ≥ 2 . Conversely, if all eigenvalues α of A have the multiplicity 1 , then all solutions of

$$Cw = 0 \qquad (w \in \mathfrak{T})$$

are in $\mathfrak{T}^+$.

Theorem 3 : The equation

$$C^2w = 0 \qquad (0 \neq w \in \mathfrak{T}^-)$$

is solvable if and only if at least one eigenvalue α of A has an order ≥ 2.

If $\dim \mathcal{R} < \infty$; then one has - this suffices for the case of different simple eigenvalues α_ν of A - in addition

$$\det C^2|_{\mathcal{J}^-} = \mathrm{discr}(A) = \mathrm{discr}\ \chi_A = \prod_{\nu<\mu} (\alpha_\nu - \alpha_\mu)^2 .$$

We remark that our above considerations can easily make Theorem 1, Theorem 2, Theorem 3 more precise by giving the dimensions. We renounce, however, an explicit formulation.

1.2.5. Applications.

At first we resume the problem, which was described in the introduction 1.2.0. in connection with the even π - periodic Mathieu functions. For the application of 1.2.3., 1.2.4., the following identification is indicated:

Let $\mathcal{R} = \mathcal{R}^* = \mathcal{T} = \mathcal{T}^*$ be the space of the even π - periodic entire analytical functions on $\mathbb{C}$. Let $\mathcal{J} = \mathcal{J}^*$ be the space of those entire functions on $\mathbb{C}^2$ which are in each variable even and π - periodic.

Moreover, define

$$\langle f,f^*\rangle := \frac{2}{\pi}\int_0^{\pi/2} f(z)f^*(z)dz$$

$$\langle h,h^*\rangle := \frac{4}{\pi^2}\int_0^{\pi/2}\int_0^{\pi/2} h(u,v)h^*(u,v)du\,dv$$

and

$$(f\otimes g)(u,v) := f(u)\cdot g(v) .$$

Then it is evident that (1) holds.

If one chooses for $\mathcal{R}$ and $\mathcal{J}$ the usual topology of uniform convergence on compact sets and if one sets

$$\langle h,f^*\rangle_1 := \frac{2}{\pi}\int_0^{\pi/2} h(u,v)f^*(v)dv ,$$

then (2) is obviously satisfied.

(3) holds true with

$$(Af)(z) := f''(z) - 2h^2\cos 2z\cdot f(z)$$

and $B = -A,\ A^* = A, B^* = B$.

For (4) one has to choose

$$[(A\otimes \mathrm{id}_{\mathcal{T}})h](u,v) := \frac{\partial^2 h}{\partial u^2}(u,v) - 2h^2\cos 2u\cdot h(u,v) ,$$

$$[(\mathrm{id}_{\mathcal{R}}\otimes B)h](u,v) := -\frac{\partial^2 h}{\partial v^2}(u,v) + 2h^2\cos 2v\cdot h(u,v)$$

and, of course, $C^* := C$.

(5) holds for normal h^2 due to MS 2.28. and more generally due to 1.1.

for arbitrary h^2 .

Finally, (6) is realized with

$$(Sh)(u,v) := h(v,u).$$

When applying the results, 1.2.3., Theorem 4 gives a generalization of a theorem on integral relations. 1.2.3., Theorem 8 yields in the special case $\gamma = 0$ the wanted expansion of a solution w of the wave equation in elliptical coordinates with the modification to 1.2.0. that ξ is to be replaced by $i\xi'$.

Now the results of 1.2.4. are applied: there every eigenvalue of A has the multiplicity 1 ; therefore 1.2.4., Theorem 2 yields:

Theorem : Every entire solution of the wave equation in elliptical coordinates

$$\frac{\partial^2 w}{\partial \xi^2} + \frac{\partial^2 w}{\partial \eta^2} + 2h^2(\mathrm{Cos}\ 2\xi - \cos 2\eta)w = 0 ,$$

which is even in both variables and πi - periodic in ξ , π - periodic in η, is symmetric:

$$w(i\eta, i\xi) = w(\xi,\eta) .$$

Finally 1.2.4., Theorem 3 furnishes the following characterization of the exceptional values:

Theorem : The exceptional values h^2 of the eigenvalues are precisely the eigenvalues h^2 of the problem

$$\left[\frac{\partial^2 \cdot}{\partial \xi^2} + \frac{\partial^2 \cdot}{\partial \eta^2} + 2h^2(\mathrm{Cos}\ 2\xi - \cos 2\eta) \cdot \right]^2 w = 0$$

in the space of the entire functions w of two variables which are even in each variable, πi - periodic in ξ and π - periodic in η , and besides antisymmetric,

$$w(i\eta, i\xi) = -w(\xi,\eta) .$$

It is known that the number of these "eigenvalues" is infinite but countable without a finite limiting point. It is also known that above every "projection point" there is only a finite number of branches; consequently every "eigenvalue" of the mentioned problem has finite multiplicity according to 1.2.4., 1.2.3., Theorem 8. We conjecture that all multiplicities are 1 , that is to say that above every "projection point" there is only a simple branching.

No entire function of one variable whose zeros are just our "eigenvalues" h^2 is known in closed form, where "closed form" means not a representation by all $a_{2n}(h^2)$ or as the limit of finite discriminants or the like.

In an analogous manner our results are applied to the eigenvalue problems MS 2.22. (II), (III), (IV) and to the eigenvalue problem with a characteristic exponent $\nu \notin \mathbb{C}$ for Mathieu's differential equation and the associated two dimensional wave equation in elliptical coordinates.

We give now an example of the application of our results to spheroidal functions by considering the characteristic eigenvalue problem to $\nu \in \mathbb{C}$ with $\nu + \frac{1}{2} \notin \mathbb{Z}$ and $\nu \pm \mu \notin \mathbb{Z}$.

Choose $\mathcal{R} = \mathcal{T}$ and $\mathcal{R}^* = \mathcal{T}^*$ as the space of the holomorphic functions f and f^*, respectively over $\mathbb{C}' := \mathbb{C} \setminus \{z : -1 \leq z \leq 1\}$ with the circuital relations $f(ze^{\pi i}) = e^{\pi i \nu} f(z)$ and $f^*(ze^{\pi i}) = e^{-\pi i(\nu+1)} f^*(z)$. Then $\mathcal{T}$ and $\mathcal{T}^*$ are the spaces of the holomorphic functions of two variables over $\mathbb{C}' \times \mathbb{C}'$ which have in both the appropiate circuital relations. Again the topology of uniform convergence on compact sets is used.

Then

$$(Af)(z) := \left[(1-z^2)f'(z)\right]' + \left[\gamma^2(1-z^2) - \frac{\mu^2}{1-z^2}\right]f(z)$$

and $B = -A$. Of course, $A \otimes \mathrm{id}_{\mathcal{T}}$ and $\mathrm{id}_{\mathcal{R}} \otimes B$ become the associated partial differential operators. Then (1) to (4) of 1.2.3. hold; (5) is already satisfied with MS 3.544. for normal γ^2 and - which is of interest in the first place - with 1.1. for arbitrary γ^2. In (6) S is again the exchange of the variables.

The results of 1.2.3. and 1.2.4. give among other things a theorem on the expansibility of solutions w over $\mathbb{C}' \times \mathbb{C}'$, with the described circuital relations, of the partial differential equation which arises from the three-dimensional wave equation in prolate spheroidal coordinates after having separated off φ. Again every such solution is symmetric. Finally, in analogy to the Mathieu case, which has been amply described above, the exceptional values of γ^2 are characterized by the corresponding quadratic eigenvalue problem; the additional remarks can be taken over mutatis mutandi.

2. Mathieu Functions.

2.1. Integral Relations.

2.1.1. Integral Relations of the first kind.

In 1972 G.Wolf has given a different derivation of the integral relations (33)-(36) in MS 2.6.. He also gave a generalization which in special cases for real parameters are related to the conditionally convergent Weber-Schafheitlin-integrals for Bessel-functions. An account of these and additional results will now be given.

Using the asymptotic behavior of the solutions of Mathieu's differential equation - see MS 2.42. (17), (18) - we ascertain that for $h \in \mathbb{C} \setminus \{0\}$ the integrals

$$(1) \qquad \frac{1}{\pi} \int_{\mathcal{L}_\rho} e^{2ihw(z,t,\alpha)}\, y_\nu(t)\,dt \qquad (\rho=3,4)$$

with

$$w(z,t,\alpha) = \operatorname{Cos} z \cos t \cos \alpha + \operatorname{Sin} z \sin t \sin \alpha ,$$

with a solution y_ν of Mathieu's differential equation to $(\lambda_\nu(h^2),h^2)$ and with the contours

$$\mathcal{L}_3 : \text{ from } -\xi + i\infty \text{ to } \eta - i\infty , \qquad (\xi,\eta \in \mathbb{R})$$
$$\mathcal{L}_4 : \text{ from } \eta - i\infty \text{ to } 2\pi - \xi + i\infty$$

are convergent in

$$z \in \mathcal{G}_\xi(\alpha) \cap \mathcal{G}_\eta(-\alpha) ,$$

where

$$\mathcal{G}_\xi(\alpha) := \{z \mid -\xi < \arg[h(\operatorname{Cos}(z + i\alpha) \pm 1)] < \pi - \xi\} .$$

Then it follows from Theorem 1 in MS 1.135. that these integrals are solutions of the modified Mathieu differential equation to $(\lambda_\nu(h^2),h^2)$.

An analogous result holds for

$$(2) \qquad \frac{1}{2\pi} \int_{\mathcal{L}_1} e^{2ihw(z,t,\alpha)}\, y_\nu(t)\,dt$$

in $\mathcal{G}_\xi(\alpha)$ with $\mathcal{L}_1$: from $-\xi + i\infty$ to $2\pi - \xi + i\infty$.

In order to identify the solutions (1) and (2) of the modified Mathieu differential equation, we introduce the entire functions $I^{(\rho)}_{y_\nu}(z;\alpha)$ $(\rho=1;3,4)$ which result from the integrals (1) and (2) by analytic continuation in z with appropriate choices of ξ,η . It is evident that

$$(3) \qquad I^{(1)}_{y_\nu} = \frac{1}{2}\left(I^{(3)}_{y_\nu} + I^{(4)}_{y_\nu}\right) .$$

For $y_\nu = me_\nu$ one verifies with

$$w(z,t,\alpha) = w(i\alpha,t,-iz)$$
$$z \in \mathcal{G}_\xi(\alpha) \Leftrightarrow i\alpha \in \mathcal{G}_\xi(-iz) \ ,$$

the identity

(4) $$I^{(1)}_{me_\nu}(z;\alpha) = I^{(1)}_{me_\nu}(i\alpha;-iz)$$

and with

$$w(z+i\pi,t,\alpha) = w(z,t-\pi,\alpha)$$
$$z + i\pi \in \mathcal{G}_\xi(\alpha) \Leftrightarrow z \in \mathcal{G}_{(\xi+\pi)}(\alpha)$$

correspondingly (substitution $t \to t+\pi$ in the integral) the identity

(5) $$I^{(1)}_{me_\nu}(z+i\pi;\alpha) = e^{i\nu\pi} I^{(1)}_{me_\nu}(z;\alpha) \ .$$

This leads with (4) and MS 2.42. (16) to

(6) $$I^{(1)}_{me_\nu}(z;\alpha) = c_\nu me_\nu(\alpha) M^{(1)}_\nu(z) \ .$$

As in (5), transformation in (1) ($t \to -t$ and $t \to 2\pi - t$) and exchange of ξ and η lead for $\nu \notin \mathbb{Z}$ to

(7) $$I^{(3)}_{me_{-\nu}}(z;\alpha) = I^{(3)}_{me_\nu}(z;-\alpha) \ .$$

(8) $$I^{(4)}_{me_{-\nu}}(z;\alpha) = e^{-2\pi i\nu} I^{(4)}_{me_\nu}(z;-\alpha) \ .$$

Together with (3) and (6) one evaluates

$$I^{(3)}_{me_\nu}(z;\alpha) = me_\nu(\alpha) \frac{e^{i\nu\frac{\pi}{2}}}{i\sin(\nu\pi)} \left\{ c_{-\nu} e^{i\nu\frac{\pi}{2}} M^{(1)}_{-\nu}(z) - c_\nu e^{-i\nu\frac{3\pi}{2}} M^{(1)}_\nu(z) \right\} \ ,$$

(9)

$$I^{(4)}_{me_\nu}(z;\alpha) = me_\nu(\alpha) \frac{-e^{i\nu\frac{\pi}{2}}}{i\sin(\nu\pi)} \left\{ c_{-\nu} e^{i\nu\frac{\pi}{2}} M^{(1)}_{-\nu}(z) - c_\nu e^{i\nu\frac{\pi}{2}} M^{(1)}_\nu(z) \right\} \ .$$

With the choice $\alpha = \xi = \eta = 0$ and $z \in \mathcal{G}_0(0)$ one has according to MS 2.68. (33)

$$I^{(3)}_{me_\nu}(z;0) = \frac{1}{\pi} \int_{i\infty}^{-i\infty} e^{2ih \operatorname{Cos} z \cos t} \, me_\nu(t)\,dt$$
$$= e^{i\nu\pi/2} me_\nu(0) M^{(3)}_\nu(z) \ .$$

Comparison with (9), making use of MS 2.42. (12), yields

$$c_\nu = e^{i\nu\pi/2} \ .$$

Therefore

$$I^{(3)}_{me_\nu}(z;\alpha) = me_\nu(\alpha)\left(e^{i\nu\frac{\pi}{2}} M^{(3)}_\nu(z)\right) \ ,$$

(10)

$$I^{(4)}_{me_\nu}(z;\alpha) = me_\nu(\alpha+\pi)\left(e^{-i\nu\frac{\pi}{2}} M^{(4)}_\nu(z)\right) \ .$$

Then it follows with MS 2.42. (11) that

$$I^{(3)}_{me_{-\nu}}(z;\alpha) = I^{(3)}_{me_\nu}(z;-\alpha) = me_{-\nu}(\alpha)\left(e^{i\nu\frac{\pi}{2}} M^{(3)}_\nu(z)\right) ,$$

and consequently that

$$I^{(3)}_{y_\nu}(z;\alpha) = y_\nu(\alpha)\left(e^{i\nu\frac{\pi}{2}} M^{(3)}_\nu(z)\right) ,$$

(11)

$$I^{(4)}_{y_\nu}(z;\alpha) = y_\nu(\alpha+\pi)\left(e^{-i\nu\frac{\pi}{2}} M^{(4)}_\nu(z)\right)$$

- in the first instance only for $\nu \in \mathbb{C} \setminus \mathbb{Z}$, but with $\nu \to n \in \mathbb{Z}$ for all $\nu \in \mathbb{C}$. Taking the derivative with respect to α yields in $\mathcal{G}_\xi(\alpha) \cap \mathcal{G}_\eta(-\alpha)$

$$\frac{1}{\pi}\int_{\mathcal{L}_3} 2ih\frac{\partial w}{\partial\alpha} e^{2ihw} y_\nu dt = y_\nu'(\alpha)e^{i\nu\frac{\pi}{2}} M^{(3)}_\nu(z) ,$$

(12)

$$\frac{1}{\pi}\int_{\mathcal{L}_4} 2ih\frac{\partial w}{\partial\alpha} e^{2ihw} y_\nu dt = y_\nu'(\alpha+\pi)e^{-i\nu\frac{\pi}{2}} M^{(4)}_\nu(z) .$$

Finally we obtain for the integral (2)

$$I^{(1)}_{y_\nu}(z;\alpha) = e^{i\nu\frac{\pi}{2}} y_\nu(\alpha)\, M^{(1)}_\nu(z) + \frac{1}{2}\left(y_\nu(\alpha+\pi)e^{-i\nu\pi} - y_\nu(\alpha)\right) e^{i\nu\frac{\pi}{2}} M^{(4)}_\nu(z) ,$$

(13)

$$\frac{1}{2\pi}\int_{\mathcal{L}_1} 2ih\frac{\partial w}{\partial\alpha} e^{2ihw} y_\nu dt = e^{i\nu\frac{\pi}{2}} y_\nu'(\alpha)\, M^{(1)}_\nu(z) + \frac{1}{2}\left(y_\nu'(\alpha+\pi)e^{-i\nu\pi} - y_\nu'(\alpha)\right) e^{i\nu\frac{\pi}{2}} M^{(4)}_\nu(z).$$

First there result all relations in MS 2.68. (35) - (39), and by specialization ($\xi = \eta = \alpha = 0$) there result from (11) and (12) the further relations in $\mathcal{G}_o(0)$

(14)

$$\int_0^\infty e^{2ih \operatorname{Cos} z \operatorname{Cos}\tau} \operatorname{Ce}_\nu(\tau)d\tau = \frac{\pi i}{2} e^{i\nu\frac{\pi}{2}} ce_\nu(0)\, M^{(3)}_\nu(z)$$

(15)

$$\int_0^\infty e^{2ih \operatorname{Cos} z \operatorname{Cos}\tau} \operatorname{Sin} z \operatorname{Sin}\tau \operatorname{Se}_\nu(\tau)d\tau = -\frac{\pi}{4h} e^{i\nu\frac{\pi}{2}} se_\nu'(0)\, M^{(3)}_\nu(z)^{*)} ,$$

and in the case of integer subscripts $m \in \mathbb{N}_o$ we have for the second solutions, also in $\mathcal{G}_o(0)$,

---

*) In the case of integer subscripts $e^{i\nu\frac{\pi}{2}} M^{(3)}_\nu$ must be replaced by $e^{-i\nu\frac{\pi}{2}} M^{(3)}_{-\nu}$.

(16) $$\int_0^\infty e^{2ih \operatorname{Cos} z \operatorname{Cos} \tau} \operatorname{Sin} z \operatorname{Sin} \tau \, Fe_m(\tau) d\tau = -\frac{\pi}{4h} i^m fe_m'(0) M_m^{(3)}(z)$$

(17) $$\int_0^\infty e^{2ih \operatorname{Cos} z \operatorname{Cos} \tau} Ge_m(\tau) d\tau = \frac{\pi}{2} i^{-m+1} ge_m(0) M_{-m}^{(3)}(z) .$$

With $h > 0$ there is now $\mathcal{G}_o(0) = \{z | \operatorname{Re} z > 0, 0 < \operatorname{Im} z < \pi\}$.

From the convergence of the integrals (14) and (15) with $\nu \in \mathbb{Z}$ and for $z > 0$ ($\operatorname{Im} z = 0$) or $0 < \operatorname{Im} z < \pi$ ($\operatorname{Re} z = 0$), respectively, follows their uniform convergence (in suitable angular domains in regard of u) and thereby the validity of (14) in $\overline{\mathcal{G}_o(0)} = \{z | \operatorname{Re} z \geq 0, 0 \leq \operatorname{Im} z \leq \pi\}$ and of (15) in $\overline{\mathcal{G}_o(0)} \setminus \{0\}$**), respectively. This is shown by the asymptotic behavior for $h > 0$ and $z \to \infty$ of the functions ($u := 2h \operatorname{Cos} z$)

$$Ce_n(z) = c_n u^{-\frac{1}{2}} \cos (u - (-1)^n \tfrac{\pi}{4}) + \mathcal{O}\left(u^{-\frac{3}{2}}\right) , \quad n \in \mathbb{N}_o$$

$$Se_n(z) = s_n u^{-\frac{1}{2}} \cos(u - (-1)^n \tfrac{\pi}{4}) + \mathcal{O}\left(u^{-\frac{3}{2}}\right) , \quad n \in \mathbb{N} .$$

- see MS <u>2.76.</u> (46) and MS <u>2.63.</u> (14) - .

By separating into real and imaginary parts and by using MS <u>2.76.</u> , there follow from (14) and (15) the following integral relations, which are valid in $n \in \mathbb{N}_o$ and with $0 < x = z < \infty$:

$$\int_0^\infty \cos(2h \operatorname{Cos} x \operatorname{Cos} \tau) \, Ce_{2n}(\tau) d\tau = \frac{\pi}{2} (-1)^{n+1} ce_{2n}(0) \, Mc_{2n}^{(2)}(x)$$

$$\int_0^\infty \sin(2h \operatorname{Cos} x \operatorname{Cos} \tau) \, Ce_{2n}(\tau) d\tau = \frac{\pi}{2} \frac{A_o^{2n}}{ce_{2n}(\frac{\pi}{2})} Ce_{2n}(x)$$

$$\int_0^\infty \cos (2h \operatorname{Cos} x \operatorname{Cos} \tau) \, Ce_{2n+1}(\tau) d\tau = \frac{\pi}{2} \frac{hA_1^{2n+1}}{ce_{2n+1}'(\frac{\pi}{2})} Ce_{2n+1}(x)$$

$$\int_0^\infty \sin(2h \operatorname{Cos} x \operatorname{Cos} \tau) \, Ce_{2n+1}(\tau) d\tau = \frac{\pi}{2} (-1)^{n+1} ce_{2n+1}(0) \, Mc_{2n+1}^{(2)}(x)$$

(18) (even for $0 \leq x < \infty$)

$$\int_0^\infty \cos(2h \operatorname{Cos} x \operatorname{Cos} \tau) \operatorname{Sin} x \operatorname{Sin} \tau \, Se_{2n+1}(\tau) d\tau = \frac{\pi}{4h} (-1)^n se_{2n+1}'(0) \, Ms_{2n+1}^{(2)}(x)$$

$$\int_0^\infty \sin(2h \operatorname{Cos} x \operatorname{Cos} \tau) \operatorname{Sin} x \operatorname{Sin} \tau \, Se_{2n+1}(\tau) d\tau = -\frac{\pi}{4} \frac{B_1^{2n+1}}{se_{2n+1}(\frac{\pi}{2})} Se_{2n+1}(x)$$

**) (14) is in $\overline{\mathcal{G}_o(0)}$ even absolutely and uniformly convergent.

$$\int_0^\infty \cos(2h \operatorname{Cos} x \operatorname{Cos} \tau) \operatorname{Sin} x \operatorname{Sin} \tau \, Se_{2n+2}(\tau)d\tau = -\frac{\pi}{2} \frac{hB_1^{2n+2}}{se'_{2n+2}(\frac{\pi}{2})} Se_{2n+2}(x)$$

$$\int_0^\infty \sin(2h \operatorname{Cos} x \operatorname{Cos} \tau) \operatorname{Sin} x \operatorname{Sin} \tau \, Se_{2n+2}(\tau)d\tau = \frac{\pi}{4h}(-1)^n se'_{2n+2}(0) \; Ms_{2n+2}^{(2)}(x)$$

In the same manner one obtains for $z = iy$

$$\int_0^\infty \cos(2h \cos y \operatorname{Cos} \tau) Ce_{2n}(\tau)d\tau = \frac{\pi}{2}(-1)^{n+1} Mc_{2n}^{(2)}(0) \; ce_{2n}(y); \; y \in \mathbb{R}$$

$$\int_0^\infty \sin(2h \cos y \operatorname{Cos} \tau) Ce_{2n}(\tau)d\tau = -\frac{\pi A_0^{2n}}{2 \cdot ce_{2n}(\frac{\pi}{2})} \left\{ ce_{2n}(y) \mp \frac{2}{\pi C_{2n}} fe_{2n}(y) \right\}$$

$$y \in \begin{cases} [0,\pi] \\ [\pi,2\pi] \end{cases}$$

$$\int_0^\infty \cos(2h \cos y \operatorname{Cos} \tau) \; Ce_{2n+1}(\tau)d\tau =$$

$$= \frac{\pi h A_1^{2n+1}}{2ce'_{2n+1}(\frac{\pi}{2})} \left\{ ce_{2n+1}(y) \mp \frac{2}{\pi C_{2n+1}} fe_{2n+1}(y) \right\} ; \; y \in \begin{cases} [0,\pi] \\ [\pi,2\pi] \end{cases}$$

$$\int_0^\infty \sin(2h \cos y \operatorname{Cos} \tau) \; Ce_{2n+1}(\tau)d\tau = \frac{\pi}{2}(-1)^{n+1} \; Mc_{2n+1}^{(2)}(0) \; ce_{2n+1}(y); \; y \in \mathbb{R}$$

(19)
$$\int_0^\infty \cos(2h \cos y \operatorname{Cos} \tau) \; \sin y \operatorname{Sin} \tau \; Se_{2n+1}(\tau)d\tau =$$

$$= \frac{\pi}{4h}(-1)^n \; Ms_{2n+1}^{(2)'}(0) \; se_{2n+1}(y) \;\; ; \; y \in (0,\pi)$$

$$\int_0^\infty \sin(2h \cos y \operatorname{Cos} \tau) \sin y \operatorname{Sin} \tau \; Se_{2n+1}(\tau)d\tau =$$

$$= -\frac{\pi B_1^{2n+1}}{4se_{2n+1}(\frac{\pi}{2})} \left\{ se_{2n+1}(y) \mp \frac{2}{\pi S_{2n+1}} ge_{2n+1}(y) \right\} ; \; y \in \begin{cases} (0,\pi) \\ (\pi,2\pi) \end{cases}$$

$$\int_0^\infty \cos(2h \cos y \operatorname{Cos} \tau) \sin y \operatorname{Sin} \tau \; Se_{2n+2}(\tau)d\tau =$$

$$= -\frac{hB_2^{2n+2}}{4se'_{2n+2}(\frac{\pi}{2})} \left\{ se_{2n+2}(y) \mp \frac{2}{\pi S_{2n+2}} ge_{2n+2}(y) \right\} ; \; y \in \begin{cases} (0,\pi) \\ (\pi,2\pi) \end{cases}$$

$$\int_0^\infty \sin(2h \cos y \operatorname{Cos} \tau) \sin y \operatorname{Sin} \tau \; Se_{2n+2}(\tau)d\tau =$$

$$= \frac{\pi}{4h}(-1)^n \; Ms_{2n+2}^{(2)'}(0) \; se_{2n+2}(y) \;\; ; \; y \in (0,\pi) \quad .$$

Completely analogous consequences of (16) and (17) are for $z = x,\ 0 < x$

$$\int_0^\infty \cos(2h \operatorname{Cos} x \operatorname{Cos} \tau)\operatorname{Sin} x \operatorname{Sin} \tau \operatorname{Fe}_{2n}(\tau)d\tau = -\frac{\pi C_{2n}\, ce_{2n}(\frac{\pi}{2})}{4hA_0^{2n}} \operatorname{Ce}_{2n}(x)$$

$$\int_0^\infty \sin(2h \operatorname{Cos} x \operatorname{Cos} \tau)\operatorname{Sin} x \operatorname{Sin} \tau \operatorname{Fe}_{2n}(\tau)d\tau = \frac{\pi}{4h}(-1)^{n+1} fe'_{2n}(0)\ Mc^{(2)}_{2n}(x)$$

$$\int_0^\infty \cos(2h \operatorname{Cos} x \operatorname{Cos} \tau)\operatorname{Sin} x \operatorname{Sin} \tau \operatorname{Fe}_{2n+1}(\tau)d\tau = \frac{\pi}{4h}(-1)^{n} fe'_{2n+1}(0)\ Mc^{(2)}_{2n+1}(x)$$

$$\int_0^\infty \sin(2h \operatorname{Cos} x \operatorname{Cos} \tau)\operatorname{Sin} x \operatorname{Sin} \tau \operatorname{Fe}_{2n+1}(\tau)d\tau = \frac{\pi C_{2n+1}\, ce'_{2n+1}(\frac{\pi}{2})}{4h^2\, A_1^{2n+1}} \operatorname{Ce}_{2n+1}(x)$$

(20)

$$\int_0^\infty \cos(2h \operatorname{Cos} x \operatorname{Cos} \tau)\operatorname{Ge}_{2n+1}(\tau)d\tau = -\frac{\pi}{2}\frac{S_{2n+1}\, se_{2n+1}(\frac{\pi}{2})}{h\ B_1^{2n+1}} \operatorname{Se}_{2n+1}(x)$$

$$\int_0^\infty \sin(2h \operatorname{Cos} x \operatorname{Cos} \tau)\operatorname{Ge}_{2n+1}(\tau)d\tau = \frac{\pi}{2}(-1)^{n+1} ge_{2n+1}(0)\ Ms^{(2)}_{2n+1}(x)$$

$$\int_0^\infty \cos(2h \operatorname{Cos} x \operatorname{Cos} \tau)\operatorname{Ge}_{2n+2}(\tau)d\tau = \frac{\pi}{2}(-1)^{n} ge_{2n+2}(0)\, Ms^{(2)}_{2n+2}(x)$$

$$\int_0^\infty \sin(2h \operatorname{Cos} x \operatorname{Cos} \tau)\operatorname{Ge}_{2n+2}(\tau)d\tau = \frac{\pi}{2}\frac{S_{2n+2}\, se'_{2n+2}(\frac{\pi}{2})}{h^2 B_2^{2n+2}} \operatorname{Se}_{2n+2}(x)$$

and for $z = iy,\ y \in (0,\pi)$:

$$\int_0^\infty \cos(2h \cos y \operatorname{Cos} \tau)\sin y \operatorname{Sin} \tau \operatorname{Fe}_{2n}(\tau)d\tau = $$
$$= (-1)^n \frac{\pi C_{2n}}{4h}\left(\frac{ce_{2n}(\frac{\pi}{2})}{A_0^{2n}}\right)^2 Mc^{(2)}_{2n}(0)\ ce_{2n}(y)$$

$$\int_0^\infty \sin(2h \cos y \operatorname{Cos} \tau)\sin y \operatorname{Sin} \tau \operatorname{Fe}_{2n}(\tau)d\tau = $$
$$= \frac{ce_{2n}(\frac{\pi}{2})}{2hA_0^{2n}}\left(\frac{\pi C_{2n}}{2} ce_{2n}(y) - fe_{2n}(y)\right)$$

$$\int_0^\infty \cos(2h \cos y \operatorname{Cos} \tau)\sin y \operatorname{Sin} \tau \operatorname{Fe}_{2n+1}(\tau)d\tau = $$
$$= \frac{ce'_{2n+1}(\frac{\pi}{2})}{2h^2 A_1^{2n+1}}\left(\frac{\pi C_{2n+1}}{2} ce_{2n+1}(y) - fe_{2n+1}(y)\right)$$

(21)

(21)

$$\int_0^\infty \sin(2h\cos y\,\mathrm{Cos}\,\tau)\sin y\,\mathrm{Sin}\,\tau\,\mathrm{Fe}_{2n+1}(\tau)d\tau =$$

$$= (-1)^{n+1}\frac{\pi C_{2n+1}}{4h}\left(\frac{ce'_{2n+1}(\frac{\pi}{2})}{hA_1^{2n+1}}\right)^2 \mathrm{Mc}_{2n+1}^{(2)}(0)\;ce_{2n+1}(y)$$

$$\int_0^\infty \cos(2h\cos y\,\mathrm{Cos}\,\tau)\mathrm{Ge}_{2n+1}(\tau)d\tau =$$

$$= \frac{\pi}{2}(-1)^n S_{2n+1}\,\mathrm{Ms}_{2n+1}^{(2)\,'}(0)\left(\frac{se_{2n+1}(\frac{\pi}{2})}{hB_1^{2n+1}}\right)^2 se_{2n+1}(y)$$

$$\int_0^\infty \sin(2h\cos y\,\mathrm{Cos}\,\tau)\mathrm{Ge}_{2n+1}(\tau)d\tau =$$

$$= -\frac{se_{2n+1}(\frac{\pi}{2})}{hB_1^{2n+1}}\left(\frac{\pi S_{2n+1}}{2}se_{2n+1}(y) - ge_{2n+1}(y)\right)$$

$$\int_0^\infty \cos(2h\cos y\,\mathrm{Cos}\,\tau)\mathrm{Ge}_{2n+2}(\tau)d\tau =$$

$$= -\frac{se'_{2n+2}(\frac{\pi}{2})}{h^2B_2^{2n+2}}\left(\frac{\pi S_{2n+2}}{2}se_{2n+2}(y) - ge_{2n+2}(y)\right)$$

$$\int_0^\infty \sin(2h\cos y\,\mathrm{Cos}\,\tau)\mathrm{Ge}_{2n+2}(\tau)d\tau =$$

$$= \frac{\pi}{2}(-1)^n S_{2n+2}\,\mathrm{Ms}_{2n+2}^{(2)\,'}(0)\left(\frac{se'_{2n+2}(\frac{\pi}{2})}{h^2B_2^{2n+2}}\right)^2 se_{2n+2}(y) \quad .$$

Further we take (13) with $\xi = \alpha = 0$ und $y_\nu = fe_m$ $(m \in \mathbb{N} \cup \{0\})$ and consider $fe'_m(0) = (-1)^m fe'_m(\pi)$. Then there results the relation

$$\frac{ih}{\pi}\int_{\mathcal{L}_1} e^{2ih\,\mathrm{Cos}\,z\cos t}\,\mathrm{Sin}\,z\sin t\;fe_m(t)dt = i^m fe'_m(0)M_m^{(1)}(z) \; .$$

For $m = 2n$ $(n \in \mathbb{N} \cup \{0\})$ we get further results after decomposition of the integration contour and after parameter transformation. The elementary evaluation yields with MS <u>2.7.</u> for $z \in \mathcal{G}_o(0)$:

$$\int_0^{\pi/2} \sin(2h\,\mathrm{Cos}\,z\cos t)\,\mathrm{Sin}\,z\sin t\left(ce_{2n}(t) - \frac{2}{\pi C_{2n}}fe_{2n}(t)\right)dt \; -$$

$$-i\int_0^{\pi/2}\cos(2h\,\mathrm{Cos}\,z\cos t)\,\mathrm{Sin}\,z\sin t\;ce_{2n}(t)dt\;-$$

$$-i\int_0^{\infty}e^{2ih\,\mathrm{Cos}\,z\,\mathrm{Cos}\,t}\,\mathrm{Sin}\,z\,\mathrm{Sin}\,t\;Ce_{2n}(t)dt=\frac{ce_{2n}(\frac{\pi}{2})}{2hA_0^{2n}}\,Ce_{2n}(z)$$

and for $m = 2n+1$

$$\int_0^{\pi/2}\cos(2h\,\mathrm{Cos}\,z\cos t)\,\mathrm{Sin}\,z\sin t\left(ce_{2n+1}(t)-\frac{2}{C_{2n+1}}fe_{2n+1}(t)\right)dt\;-$$

$$-i\int_0^{\pi/2}\sin(2h\,\mathrm{Cos}\,z\cos t)\,\mathrm{Sin}\,z\sin t\;ce_{2n+1}(t)dt\;+$$

$$+\int_0^{\infty}e^{2ih\,\mathrm{Cos}\,z\cos t}\,\mathrm{Sin}\,z\,\mathrm{Sin}\,t\;Ce_{2n+1}(t)dt=\frac{ce'_{2n+1}(\frac{\pi}{2})}{4h^2\,A_1^{2n+1}}\,Ce_{2n+1}(z)$$

The corresponding consideration with (13) and $y_\nu = ge_m$ yields

$$i\int_0^{\pi/2}\sin(2h\,\mathrm{Cos}\,z\cos t)\left(se_{2n+1}(t)-\frac{2}{\pi S_{2n+1}}ge_{2n+1}(t)\right)dt\;+$$

$$+\int_0^{\pi/2}\cos(2h\,\mathrm{Cos}\,z\cos t)\;se_{2n+1}(t)dt\;+$$

$$+\int_0^{\infty}e^{2ih\,\mathrm{Cos}\,z\,\mathrm{Cos}\,t}\,Se_{2n+1}(t)dt\;=$$

$$=\;(-1)^n\,\frac{se_{2n+1}(\frac{\pi}{2})}{hB_1^{2n+1}}\,Se_{2n+1}(z)$$

and

$$\int_0^{\pi/2}\cos(2h\,\mathrm{Cos}\,z\cos t)\left(se_{2n+2}(t)-\frac{2}{S_{2n+2}}ge_{2n+2}(t)\right)dt\;+$$

$$+i\int_0^{\pi/2}\sin(2h\,\mathrm{Cos}\,z\cos t)\;se_{2n+2}(t)dt\;+$$

$$+\int_0^{\infty}e^{2ih\,\mathrm{Cos}\,z\,\mathrm{Cos}\,t}\,Se_{2n+2}(t)dt\;=$$

$$=\;(-1)^n\,\frac{se'_{2n+2}(\frac{\pi}{2})}{h^2B_2^{2n+2}}\,Se_{2n+2}(z)$$

(valid in each case for $n \in \mathbb{N} \cup \{0\}$ and $z \in \mathcal{G}_o(0)$).

Another 16 new real relations can be gained by decomposition into real and imaginary parts for $z = x \in (0,\infty)$ and for $z = iy$ with $y \in (0,\pi)$. However, we abstain from noting them explicitly.

2.1.2. Integral Relations of the second kind (with variable boundaries).

We start from the (time independent) two dimensional wave equation in elliptical coordinates (ξ,η)

$$u_{\xi\xi} + u_{\eta\eta} + 2h^2(\mathrm{Cos}\ 2\xi - \cos 2\eta)u = 0 ,$$

which will be considered in $\mathbb{C}^2$ for $h^2 \in \mathbb{C} \setminus \{0\}$. We substitute $\xi \longrightarrow i\xi$ and maintain the function symbol u ; then

$$(1) \qquad \partial_2^2 u - \partial_1^2 u + 2h^2(\cos 2\xi - \cos 2\eta)u = 0 .$$

Here and in the following ∂_1 and ∂_2 denote the partial derivatives with regard to the first and to the second variable, respectively, which in (1) are ξ,η. We introduce now the variables

$$(2) \qquad x = \xi + \eta,\ y = \eta - \xi\ ,\ u(\xi,\eta) = w(x,y)$$

with which (1) is transformed into

$$(3) \qquad \partial_1\partial_2 w + h^2 \sin x \sin y\ w = 0 .$$

One recognizes a close relation to the telegraphists' equation (4) :

<u>Lemma :</u> If $\psi \in C^2$ is a solution of

$$(4) \qquad \partial_1\partial_2\psi + \psi = 0 ,$$

then $w(x,y) = \psi(h \cos x,\ h \cos y)$ is a solution of (3).

We renounce a discussion of the corresponding local transformation.

It is known and easily verifiable that for $(\alpha_o,\beta_o) \in \mathbb{C}^2$

$$\psi(\alpha,\beta) := J_o(2\sqrt{(\alpha-\alpha_o)(\beta-\beta_o)})$$

is an entire solution of (4) ; in fact ψ is its <u>Riemann function</u>: one has $\psi(\alpha,\beta_o) = \psi(\alpha_o,\beta) = 1$.

Then the given transformation gives the Riemann function of (3)

$$(5) \qquad v(x,y) := J_o(2h\sqrt{(\cos x - \cos x_o)(\cos y - \cos y_o)})$$

for the point $(x_o,y_o) \in \mathbb{C}^2$.

We apply now Riemann's integration method with (5) to the differential equation (3) which is formally of the hyperbolic type. A short proof will now be given.

Let w be an entire solution of (3) and let v be the entire solution (5). Then with

$$f := v\partial_1 w\ ,\ g := w\partial_2 v$$

one obviously has

$$\partial_2 f = \partial_1 g .$$

Therefore, for instance,

$$F(x,y) := \int_0^x f(\xi,0)\,d\xi + \int_0^y g(x,\eta)\,d\eta$$

furnishes an entire primitive function to (f,g). Therefore we have

$$\int_{\mathcal{L}} (f(x,y)dx + g(x,y)dy) = 0$$

for a closed continuous contour $\mathcal{L}$ in $\mathbb{C}^2$ which is piecewise continuously differentiable. We choose in particular the triangular contour

$$(x_o,y_o) \longrightarrow (y_o,y_o) \longrightarrow (x_o,x_o) \longrightarrow (x_o,y_o) .$$

On the third leg of this triangle x is constant, and so is v (= 1) ; equally on the first leg y is constant and $v = 1$. Thus one obtains at once

$$\text{(6)} \qquad w(x_o,y_o) = w(y_o,y_o) + \int_{y_o}^{x_o} (v\partial_1 w + w\partial_2 v)(x,x)\,dx .$$

Now with (2) we go back to (1) using the obvious relation

$$\partial_1 w = \frac{1}{2}\partial_1 u + \frac{1}{2}\partial_2 u .$$

Thus one obtains

$$\text{(7)} \qquad \begin{aligned} u(\xi,\eta) &= u(0,\eta-\xi) + \int_{\eta-\xi}^{\eta+\xi} u(0,z)k_2(\xi,\eta,z)\,dz + \\ &+ \frac{1}{2}\int_{\eta-\xi}^{\eta+\xi} (\partial_1 u(0,z) + \partial_2 u(0,z))k_1(\xi,\eta,z)\,dz \end{aligned}$$

with

$$\text{(8)} \qquad k_1(\xi,\eta,z) := J_o(2h\sqrt{(\cos(\xi+\eta) - \cos z)(\cos(\eta-\xi) - \cos z)})$$

and

$$\text{(9)} \qquad k_2(\xi,\eta,z) := 2h^2 \sin z\, \frac{\cos z - \cos(\xi+\eta)}{2h\sqrt{(\ldots)(\ldots)}}\, J_1(2h\sqrt{(\ldots)(\ldots)}).$$

Let y_1, y_2 be, in particular, two arbitrary solutions of Mathieu's differential equation to the same parameter pair (λ,h^2) , then one can choose $u(\xi,\eta) = y_1(\xi)y_2(\eta)$ and obtains

$$\text{(10)} \qquad \begin{aligned} y_1(\xi)y_2(\eta) &= y_1(0)y_2(\eta-\xi) + \int_{\eta-\xi}^{\eta+\xi} y_1(0)y_2(z)k_2(\xi,\eta,z)\,dz + \\ &+ \frac{1}{2}\int_{\eta-\xi}^{\eta+\xi} (y_1'(0)y_2(z) + y_1(0)y_2'(z))k_1(\xi,\eta,z)\,dz . \end{aligned}$$

An important and interesting special case is obtained with $y_1 = y_{II}$, that is with $y_1(0) = 0,\ y_1'(0) = 1$:

$$2y_{II}(\xi)y(\eta) = \int_{\eta-\xi}^{\eta+\xi} k_1(\xi,\eta,z)y(z)dz \ , \tag{11}$$

where $y(\eta)$ is still an arbitrary solution of Mathieu's differential equation. It is verified by a simple calculation that k_1 is a symmetric function of ξ,η,z. Therefore (10) is seen to be the integral relation of MS 2.82. for $s = 0$. We recognize here why just $s = 0$ occupies a special position: in this case we have just the Riemann function.

We point to a few consequences: If y is π - periodic and even, then for $\xi = \frac{\pi}{2}$ also the integrand in (11) is π - periodic and even and one obtains with fixed bounds -

$$\int_0^{\pi/2} J_0\left(2h\sqrt{\cos^2 z - \sin^2\eta}\right)ce_{2n}z dz = y_{II}\left(\frac{\pi}{2};a_{2n},h^2\right)\ ce_{2n}\eta \ . \tag{12}$$

If y is 2π - periodic and even, then for $\xi = \pi$ the integrand in (11) becomes 2π - periodic and even and one obtains

$$\int_0^{\pi/2} J_0(2h(\cos z + \cos\eta))ce_n z dz = y_{II}(\pi;a_n,h^2)ce_n\eta \ . \tag{13}$$

Of course, (12) and (13) are in principle contained in the formulas and theorems of MS. However, the representation of the eigenvalues of the integral equations (12) and (13), which is here obtained, appears to be remarkable.

2.2. Addition theorems.

As a supplement to the addition theorems in MS 2.5., we give here addition theorems which are based on the transformation

$$c_0\ \mathrm{Cos}(z_0 \pm it_0) = e^{\pm i\alpha}c\ \mathrm{Cos}(z \pm it) + e^{\pm i\beta}\ \gamma\mathrm{Cos}(\zeta \pm i\tau) \ . \tag{1}$$

It generalizes the transformation in MS 2.52. Lemma 1 and leads to a generalization of the expansions given there, with expansion coefficients being given by series in Mathieu functions instead of series in cylinder functions. When solving (1) for z_0,t_0, there result two (z,t) - domains $\mathcal{B}_e$ and $\mathcal{B}_i$, for each of which we obtain an addition theorem (exterior and interior addition theorem, respectively). By (1), each addition theorem follows from the other one by a certain rearrangement of the terms of the series. All known addition theorems, those in MS 2.5. as well as those of K.Saermark (1959), K.Germey (1964) and G.Wolf (1968), turn out to be special cases as a consequence of degeneracies of the coordinate systems in (1) . In the explicit demonstration we lean against G.Wolf 1969, utilizing the principles of MS 2.51..

2.2.1. Lemmas concerning the transformation equation.

For complex parameters $c(\neq 0)$, $c_o(\neq 0)$, $\gamma(\neq 0)$, α,β,ζ and τ we designate - in correspondence to MS 2.52. -

$$(2)\qquad \begin{aligned} A^+ &= \max(\mathrm{Re}\ \zeta_1,\ \mathrm{Re}\ \zeta_2) \\ A^- &= \max(\mathrm{Re}\ \zeta_3,\ \mathrm{Re}\ \zeta_4) \\ \mathscr{B}_e &= \{(z,t) \mid \mathrm{Re}\ z - \mathrm{Im}\ t > A^+,\ \mathrm{Re}\ z + \mathrm{Im}\ t > A^-\} \end{aligned}$$

and

$$(3)\qquad \begin{aligned} B^+ &= \min(\mathrm{Re}\ \zeta_1,\ \mathrm{Re}\ \zeta_2) \\ B^- &= \min(\mathrm{Re}\ \zeta_3,\ \mathrm{Re}\ \zeta_4) \\ \mathscr{B}_i &= \{(z,t) \mid |\mathrm{Re}\ z - \mathrm{Im}\ t| < B^+,\ |\mathrm{Re}\ z + \mathrm{Im}\ t| < B^-\} \end{aligned}$$

where the ζ_κ with $\mathrm{Re}\ \zeta_\kappa \geqq 0$ $(\kappa \in \{1,2,3,4\})$ are uniquely determined mod $2\pi i$ by

$$\begin{aligned} e^{i\alpha}c\ \mathrm{Cos}\ \zeta_1 &= c_0 - e^{i\beta}\gamma\ \mathrm{Cos}(\zeta + i\tau) \\ e^{i\alpha}c\ \mathrm{Cos}\ \zeta_2 &= -c_0 - e^{i\beta}\gamma\ \mathrm{Cos}(\zeta + i\tau) \\ e^{-i\alpha}c\ \mathrm{Cos}\ \zeta_3 &= c_0 - e^{-i\beta}\gamma\ \mathrm{Cos}(\zeta - i\tau) \\ e^{-i\alpha}c\ \mathrm{Cos}\ \zeta_4 &= -c_0 - e^{-i\beta}\gamma\ \mathrm{Cos}(\zeta - i\tau) \end{aligned}$$

Then there holds

Lemma 1 :

a) There exist two functions z_0 and t_0 which are holomorphic in $\mathscr{B}_e$ and solve the equations

$$(1)\quad c_0\ \mathrm{Cos}(z_0 \pm it_0) = e^{\pm i\alpha}c\ \mathrm{Cos}(z \pm it) + e^{\pm i\beta}\gamma\ \mathrm{Cos}(\zeta \pm i\tau)\ ,$$

and satisfy $\mathrm{Re}(z_0 \pm it_0) > 0$ and

$$(4)\qquad \begin{aligned} z_0(z,t+2\pi) &= z_0(z,t) \\ t_0(z,t+2\pi) &= t_0(z,t) + 2\pi \\ z_0(z+2\pi i,t) &= z_0(z,t) + 2\pi i \\ t_0(z+2\pi i,t) &= t_0(z,t)\ . \end{aligned}$$

The functions $z_0 \pm it_0$ are uniquely determined mod $2\pi i$.

b) For

$$C^{\pm} = \min_{\sigma \in [-1,1]} \left\{\mathrm{Re}\ \omega^{\pm}(\sigma) \mid e^{\pm i\alpha}c\ \mathrm{Cos}\ \omega^{\pm}(\sigma) = c_0\sigma - e^{\pm i\beta}\gamma\ \mathrm{Cos}(\zeta \pm i\tau);\ \mathrm{Re}\ \omega^{\pm}(\sigma) \geqq 0\right\} > 0^{*)}$$

---

*) For real parameters $c(>0)$, $c_o(>0)$, $\gamma(>0)$, α,β,ζ and τ a geometrical interpretation is that the intervals $\overline{-c_0, c_0}$ and $\overline{-e^{i\alpha}c,\ e^{i\alpha}c}$ do not intersect.

- which, because of $B^+ \geqq C^+$ and $B^- \geqq C^-$, entails $\mathcal{B}_i \neq \emptyset$ - there exist two in $\mathcal{B}_i$ holomorphic solutions of (1) : $z_o = \hat{z}_o$ and $t_o = \hat{t}_o$ with $Re(\hat{z}_o(0,0) \pm i\hat{t}_o(0,0)) > 0$. $\hat{z}_o$ and $\hat{t}_o$ are $2\pi i$ - periodic in z and 2π - periodic in t ; with respect to $\hat{z}_o \pm i\hat{t}_o$ they are uniquely determined mod $2\pi i$.

c) If v is a solution of

$$(5)\quad \frac{\partial^2 v}{\partial z_o^2} + 2h_o^2 \operatorname{Cos} 2z_o \cdot v = -\frac{\partial^2 v}{\partial t_o^2} + 2h_o^2 \cos 2t_o \cdot v$$

with $h_o = \frac{1}{2} kc_o$ $\qquad$ ($k \in \mathbb{C}$ arbitrary) ,
which is an entire holomorphic function of z_o and t_o , then the functions

$$u_e(z,t) = v(z_o,t_o) \quad \text{and} \quad u_i(z,t) = v(\hat{z}_o,\hat{t}_o)$$

which are holomorphic functions in $\mathcal{B}_e$ and in $\mathcal{B}_i$, respectively, satisfy

$$\frac{\partial^2 u}{\partial z^2} + 2h^2 \operatorname{Cos} 2z \cdot u = -\frac{\partial^2 u}{\partial t^2} + 2h^2 \cos 2t \cdot u \quad \text{with } h = \frac{1}{2} kc .$$

The proofs of a) and b) follow in a simple manner by a detailed investigation of the function Cos and of its inverse function. The proof of c) is similar to the one in MS 2.52..

Besides the known asymptotic behavior of Mathieu functions, the following lemma is of great importance in the derivation of the integral relations:

<u>Lemma 2 :</u> For $Re\, z \to \infty$ there holds, with a suitable fixation of $z_o \pm it_o$ mod $2\pi i$, with the abbreviation

$$w(x,\chi,\psi) = \frac{1}{2}(e^x \cos(\chi-\psi) + e^{-x}\cos(\chi+\psi)) \qquad ((x;\chi,\psi) \in \mathbb{C}^3)$$

and bounded $Im\, t$, uniformly

$$2h_o \operatorname{Cos} z_o(z,t) = 2h \operatorname{Cos} z + 2\hat{h}\, w(\zeta;\tau,t+\alpha-\beta) + \mathcal{O}\left(\frac{1}{\operatorname{Cos} z}\right) ,$$

$$t_o(z,t) = t + \alpha + \mathcal{O}\left(\frac{1}{\operatorname{Cos} z}\right) ,$$

$$\arg(2h \operatorname{Cos} z) \sim \arg(2h_o \operatorname{Cos} z_o) .$$

Besides, one has to set $h_o = \frac{1}{2} kc_o$, $h = \frac{1}{2} kc$ and $\hat{h} = \frac{1}{2} k\gamma$ with arbitrary k .

Division and multiplication of the two equations (1) and the asymptotic behavior of elementary functions yield the proof of the lemma.-

2.2.2. Integral relations.

In the notations and in the presuppositions we follow the Lemmas 1 and 2. Then there holds in addition

<u>Theorem 1 :</u> For $j \in \{1,2,3,4\}$ and $\nu \in \mathbb{C}$ we introduce the functions

$$u_\nu^{(j)} = M_\nu^{(j)}(z_o;h_o) \cdot me_\nu(t_o;h_o^2) ,$$

which are holomorphic in $\mathcal{B}_e$. Then one has for normal values to 0 and 1 of

$$\hat{h}^2 \left(\hat{h} = \frac{1}{2} k\gamma\right) \qquad \text{and for all integers } n \text{ in}$$

$$\mathrm{Re}\, z > \max(A^+ + \mathrm{Im}\,\theta,\ A^- - \mathrm{Im}\,\theta) \qquad (\theta \in \mathbb{C})$$

the expansion

$$\frac{1}{2\pi} \int_\theta^{\theta+2\pi} u_\nu^{(j)}(z,t) me_{\nu+n}(-t;h^2)dt = \left\{ \sum_{m=-\infty}^{\infty} K_{mn} M_m^{(1)}(\zeta,\hat{h}) me_m(\tau,\hat{h}^2) \right\} \cdot M_{\nu+n}^{(j)}(z;h)$$

with

$$K_{mn}(\alpha,h^2;\beta,\hat{h}^2) = i^{m+n} \frac{1}{2\pi} \int_0^{2\pi} me_m(-t+\beta;\hat{h}^2) me_\nu(t;h_o^2) me_{\nu+n}(-t+\alpha;h^2)dt ,$$

where $K_{mn} = 0$ for $m + n = \text{odd}$.

For the proof we choose in Lemma 1c)

$$v(z_o,t_o) = M_\nu^{(j)}(z_o;h_o) me_\nu(t_o;h_o^2) .$$

Then the functions $u_\nu^{(j)}$ are solutions of (6) with

$$u_\nu^{(j)}(z,t+2\pi) = e^{2\pi i\nu} u_\nu^{(j)}(z,t) .$$

This leads with MS 1.135. to the solutions

$$I_n^{(j)}(z) = \frac{1}{2\pi} \int_\theta^{\theta+2\pi} u_\nu^{(j)}(z,t) me_{\nu+n}(-t;h^2)dt$$

of the modified Mathieu differential equation to the parameter pair $\left(\lambda_{\nu+n}(h^2),h^2\right)$. An examination of the asymptotic behavior of $I_n^{(j)}$ permits to identify this solution. For $j = 3$, with Lemma 2 and with the asymptotic formulas for $M_\nu^{(3)}$ and $\mathfrak{z}_\nu^{(3)} = H_\nu^{(1)}$, we have in

$$\mathrm{Re}\, z > \max(A^+ + \mathrm{Im}\,\theta,\ A^- - \mathrm{Im}\,\theta)$$

the relations

$$I_n^{(3)}(z) = \frac{1}{2\pi} \int_\theta^{\theta+2\pi} u_\nu^{(3)}(z,t) me_{\nu+n}(-t;h^2)dt$$

$$= \frac{1}{2\pi} \int_\theta^{\theta+2\pi} e^{2i\hat{h}w(\zeta;\tau,t+\alpha-\beta)} me_\nu(t+\alpha;h_o^2)\ me_{\nu+n}(-t;h^2)dt \cdot$$

$$\cdot\ i^n \mathfrak{z}_{\nu+n}^{(3)}(2h \operatorname{Cos} z) \left(1 + \mathcal{O}\left(\frac{1}{\operatorname{Cos} z}\right)\right)$$

$$= \frac{i^n}{2\pi} \int_0^{2\pi} e^{2i\hat{h}w(\zeta;\tau,t+\alpha-\beta)} me_\nu(t+\alpha;h_o^2) me_{\nu+n}(-t,h^2)dt \cdot M_{\nu+n}^{(3)}(z;h) .$$

With MS 2.68. (38) and after interchange of summation and integration there results

$I_n^{(3)}(z) =$

$$= M_{\nu+n}^{(3)}(z;h) \sum_{m=-\infty}^{\infty} i^{n+m} \frac{1}{2\pi} \int_0^{2\pi} me_m(-t-\alpha+\beta;\hat{h}^2) me_\nu(t+\alpha;h_o^2) me_{\nu+n}(-t;h^2)dt \cdot M_m^{(1)}(\zeta;\hat{h}) me_m(\tau;\hat{h}^2) .$$

This is just the assertion for $j = 3$ if in the integral t is substituted for $t + \alpha$. Taking notice of $K_{mn} = 0$ for $n + m = \text{odd}$, the corresponding result for $j = 4$ is obtained.

Finally we record the representation of the K_{mn} by means of the Fourier coefficients of me_ν :

$$K_{2\ell-n,n} = (-1)^\ell \sum_{p,q=-\infty}^{\infty} c_{2p-2\ell}^{2\ell-n}(\hat{h}^2)\, e^{i(2p-n)\beta} c_{2q}^{\nu+n}(h^2)\, e^{i(\nu+n+2q)\alpha} c_{2p+2q}^{\nu}(h_o^2) .$$

2.2.3. The addition theorems.

Just as in MS 2.54. one concludes now with Theorem 1 to

Theorem 2 : (exterior addition theorem):

Let h^2 be normal value to ν and $\nu + 1$ $(\nu \in \mathbb{C})$ and let $\hat{h}^2$ be normal value to 0 and 1. Then, with the notations and presuppositions of MS 2.921., there holds in $\mathcal{B}_e$ for non-integer ν :

$$M_\nu^{(j)}(z_o(z,t);h_o) me_\nu(t_o(z,t);h_o^2) =$$

$$(7) \qquad = \sum_{n=-\infty}^{\infty} \left\{ \sum_{m=-\infty}^{\infty} K_{mn} M_m^{(1)}(\zeta;\hat{h}) me_m(\tau;\hat{h}^2) \right\} M_{\nu+n}^{(j)}(z;h) me_{\nu+n}(t;h^2)$$

with K_{mn} from Theorem 1 ; and there holds for integer ν :

$$M_\nu^{(j)}(z_o(z,t);h_o) me_\nu(t_o(z,t);h_o^2) = \left\{ \sum_{m=-\infty}^{\infty} K_{mo} M_m^{(1)}(\zeta;\hat{h}) me_m(\tau;\hat{h}^2) \right\} M_o^{(j)}(z;h) me_o(t;h^2) +$$

$$(8) \qquad + \sum_{n=1}^{\infty} \left[\left\{ \sum_{m=-\infty}^{\infty} K_{mn} M_m^{(1)}(\zeta;\hat{h}) me_m(\tau;\hat{h}^2) \right\} M_n^{(j)}(z;h) me_n(t;h^2) + \right.$$

$$\left. + \left\{ \sum_{m=-\infty}^{\infty} K_{m,-n} M_m^{(1)}(\zeta;\hat{h}) me_m(\tau;\hat{h}^2) \right\} M_{-n}^{(j)}(z;h) me_{-n}(t;h^2) \right]$$

with $$K_{mn} = i^{m+n} \frac{1}{2\pi} \int_0^{2\pi} me_m(-t+\beta;\hat{h}^2) me_\nu(t;h_o^2) me_n(-t+\alpha;h^2)dt .^{*)}$$

Interchange of the parameters α, h^2, z, t with $\beta, \hat{h}^2, \zeta, \tau$ and rearrangement of the series yields formally the corresponding

*) For $\text{Re}\, z \pm \text{Im}\, t > A = \max(A^+, A^-)$ - in particular for $A^+ = A^-$ in $\mathcal{B}_e$ - there holds besides (8) also the expansion (7).

Theorem 2 : (interior addition theorem):

Let h^2 be normal value to 0 and 1 and let $\hat{h}^2$ be normal value to ν and $\nu+1$. With $\mathrm{Re}(\zeta \pm i\tau) > 0$ and

$$\min_{\varphi \in [0,2\pi]} \left\{ \min_{\sigma\in[-1,1]} |e^{\pm i\beta_\gamma} \mathrm{Cos}(\zeta \pm i\tau + i\varphi) - e^{\pm i\alpha} c\sigma| \right\} > |c_o| \quad (9) \; {}^{*)}$$

and with the statements of Lemma 1 on $\hat{z}_o$ and $\hat{t}_o$ there holds in $\mathcal{B}_1$

$$\begin{aligned} M_\nu^{(j)}(\hat{z}_o(z,t);h_o)\mathrm{me}_\nu\left(\hat{t}_o(z,t);h_o^2\right) = \\ = \left\{ \sum_{m=-\infty}^{\infty} L_{mo} M_{\nu+m}^{(j)}(\zeta;\hat{h})\mathrm{me}_{\nu+m}(\tau;\hat{h}^2) \right\} \cdot M_o^{(1)}(z;h)\mathrm{me}_o(t;h^2) + \\ + \sum_{n=1}^{\infty} \left[\left\{ \sum_{m=\infty}^{\infty} L_{mn} M_{\nu+m}^{(j)}(\zeta;\hat{h})\mathrm{me}_{\nu+m}(\tau;\hat{h}^2) \right\} M_n^{(1)}(z;h)\mathrm{me}_n(t;h^2) + \right. \\ \left. + \left\{ \sum_{m=-\infty}^{\infty} L_{m,-n} M_{\nu+m}^{(j)}(\zeta;\hat{h})\mathrm{me}_{\nu+m}(\tau;\hat{h}^2) \right\} M_{-n}^{(1)}(z;h)\mathrm{me}_{-n}(t;h^2) \right] \end{aligned} \quad (10)$$

and in $|\mathrm{Re}\, z \pm \mathrm{Im}\, t| < B = \min(B^+,B^-)$ there holds

$$\begin{aligned} M_\nu^{(j)}(\hat{z}_o(z,t);h_o)\mathrm{me}_\nu\left(\hat{t}_o(z,t);h_o^2\right) = \\ = \sum_{n=-\infty}^{\infty} \left\{ \sum_{m=-\infty}^{\infty} L_{mn} M_{\nu+m}^{(j)}(\zeta;\hat{h})\mathrm{me}_{\nu+m}(\tau;\hat{h}^2) \right\} M_n^{(1)}(z;h)\mathrm{me}_n(t;h^2) \end{aligned}$$

with the constants from Theorem 1

$$L_{mn} = K_{nm}(\beta,\hat{h}^2;\alpha,h^2) \quad .$$

For the proof we expand at first for every z with $|\mathrm{Re}\, z| < \frac{1}{2}(B^+ + B^-)$ the function of t

$$\hat{u}_\nu^{(j)}(z,t) = M_\nu^{(j)}\left(\hat{z}_o(z,t);h_o\right)\mathrm{me}_\nu\left(\hat{t}_o(z,t);h_o^2\right)$$

in the stripe

$$\max(-B^+ + \mathrm{Re}\, z, -B^- - \mathrm{Re}\, z) < \mathrm{Im}\, t < \min(B^+ + \mathrm{Re}\, z, B^- - \mathrm{Re}\, z)$$

as

$$\hat{u}_\nu^{(j)}(z,t) = b_o^{(j)}(z)\mathrm{me}_o(t;h^2) + \sum_{n=1}^{\infty} \left\{ b_n^{(j)}(z)\mathrm{me}_n(t;h^2) + b_{-n}^{(j)}(z)\mathrm{me}_{-n}(t;h^2) \right\} \quad (+)$$

which is possible since

$$\hat{u}_\nu^{(j)}(z,t+2\pi) = \hat{u}_\nu^{(j)}(z,t) \quad .$$

---

*) By suitable choice of β,γ and (ζ,τ) in $e^{\pm i\beta_\gamma} \mathrm{Cos}(\zeta \pm i\tau) = \mathrm{const}$, (9) can be replaced by

$$\min_{\sigma \in [-1,1]} \left\{ |e^{\pm i\beta_\gamma} \mathrm{Cos}(\zeta \pm i\tau) - e^{\pm i\alpha} c\sigma| \right\} > |c_o|$$

Then the coefficients

$$b_n^{(j)}(z) = \frac{1}{2\pi} \int_{\theta}^{\theta+2\pi} \hat{u}_\nu^{(j)}(z,t) me_n(-t;h^2)dt$$

are in

$$\max(-B^+ + \operatorname{Im}\theta, -B^- - \operatorname{Im}\theta) < z < \min(B^+ + \operatorname{Im}\theta, B^- - \operatorname{Im}\theta)$$

solutions of the modified Mathieu differential equation to $(\lambda_n(h^2),h^2)$. Then by Lemma 1 there is $b_n^{(j)}(z+2\pi i) = b_n^{(j)}(z)$, which entails

$$b_n^{(j)}(z) = B_n^{(j)} \; M_n^{(1)}(z;h) \; .$$

For the evaluation of the $B_n^{(j)}$ we examine their dependence on (ζ,τ). For that we choose according to (9) $\delta > 0$ and $D^{\pm} > 0$, such that with

$$\mathcal{B}_\delta := \{(z,t) \mid |\operatorname{Re}(z \pm it)| < \delta\}$$

and

$$\vartheta := \{(\hat{\zeta},\hat{\tau}) \mid \operatorname{Re}(\hat{\zeta} \pm i\hat{\tau}) > D^{\pm}\}$$

$$(\zeta,\tau) \in \vartheta \text{ and for } (\hat{\zeta},\hat{\tau}) \in \vartheta$$

there always holds

$$\min_{\varphi\in[0,2\pi]} \left\{ \inf_{(z,t)\in\mathcal{B}_\delta} |e^{\pm i\beta}\gamma \operatorname{Cos}(\hat{\zeta} \pm i\hat{\tau} + i\varphi) - e^{\pm i\alpha}c \operatorname{Cos}(z \pm it)| \right\} > |c_0| \quad .$$

Therefore we can solve the transformation equation (1) - (ζ,τ) is there to be replaced by $(\hat{\zeta},\hat{\tau})$ - for functions $\tilde{z}_o$ and $\tilde{t}_o$ (for z_o,t_o) , which are holomorphic in $\mathcal{B}_\delta \times \vartheta$. Their properties can be read off from Lemma 1a and Lemma 1b. In particular we can, with respect to $(\hat{\zeta},\hat{\tau}) \in \vartheta$, apply Theorem 2 $(\nu \in \mathbb{C} \setminus \mathbb{Z})$ to

$$\tilde{u}_\nu^{(j)}(z,t;\hat{\zeta},\hat{\tau}) = M_\nu^{(j)}(\tilde{z}_o(z,t;\hat{\zeta},\hat{\tau});h_o) me_\nu(\tilde{t}_o(z,t;\hat{\zeta},\hat{\tau});h_o^2)$$

for every $(z,t) \in \mathcal{B}_\vartheta$, and we obtain

$$\tilde{u}_\nu^{(j)}(z,t;\hat{\zeta},\hat{\tau}) = \sum_{m=-\infty}^{\infty} \left\{ \sum_{n=-\infty}^{\infty} L_{mn} M_n^{(1)}(z;h) me_n(t;h^2) \right\} M_{\nu+m}^{(j)}(\hat{\zeta};\hat{h}) me_{\nu+m}(\hat{\tau};\hat{h}^2)$$

with $L_{m,n} = K_{n,m}(\beta,\hat{h}^2;\alpha,h^2)$ (see Theorem 1).

Now, because of its absolute convergence, this series can for $\hat{\zeta} = \zeta$ and for every $z(|\operatorname{Re} z| < \delta)$
in $\{t \mid |\operatorname{Im} t| < \delta - |\operatorname{Re} z|\} \times \{\hat{\tau} | \operatorname{Re}\zeta - D^+ > \operatorname{Im}\hat{\tau} > -\operatorname{Re}\zeta + D^-\}$

be rearranged into

$$\tilde{u}_\nu^{(j)}(z,t;\zeta,\hat{\tau}) = \sum_{n=-\infty}^{\infty} \left\{ \sum_{m=-\infty}^{\infty} L_{mn} M_{\nu+m}^{(j)}(\zeta;\hat{h}^2) me_{\nu+m}(\hat{\tau};\hat{h}^2) \right\} M_n^{(1)}(z;h) me_n(t;h^2) \; .$$

For $\hat{\tau} = \tau$ we compare with (+) , and thus we obtain

$$B_n^{(j)} = \sum_{m=-\infty}^{\infty} L_{mn} M_{\nu+m}^{(j)}(\zeta;\hat{h}) me_{\nu+m}(\tau;\hat{h}^2) \; .$$

In the case of integer ν and $|\mathrm{Im}\,\tau| \geq \min(\mathrm{Re}\,\zeta - D^+,\ \mathrm{Re}\,\zeta + D^+)$, the terms in the series with $\nu + m$ and $-\nu - m$ must be combined.

2.2.4. Consequences and special cases.

At first we consider the following degeneracies of the elliptic coordinate system:

$$\gamma\ \mathrm{Cos}(\zeta \pm i\tau) = \rho e^{\pm i\psi} \quad \text{with} \quad \beta = 0,\ \gamma \to 0,\ 2\hat{h}\ \mathrm{Cos}\,\zeta \to k\rho,\ \tau \to \psi \ .$$

It leads to the addition theorem MS 2.54. Theorem 2 as well as to the corresponding interior addition theorem (see G.Wolf 1968)

$$M_\nu^{(j)}(z_o,h_o)me_\nu(t_o;h_o^2) = B_o^{(j)}M_o^{(1)}(z,h)me_o(t;h^2) +$$

$$+ \sum_{s=1}^{\infty}\left\{B_s^{(j)}M_s^{(1)}(z;h)me_s(t;h^2) + B_{-s}^{(j)}M_{-s}^{(1)}(z;h)me_{-s}(t;h^2)\right\}$$

$$\left(j = 1,2,3,4;\ h = \tfrac{1}{2}\,kc;\ h_o = \tfrac{1}{2}kc_o\right)$$

with the absolutely convergent series

$$B_s^{(j)} = \sum_{\ell=-\infty}^{\infty}(-1)^\ell \sum_{p=-\infty}^{\infty} c_{2p}^{\nu}(h_o^2)c_{2p-2\ell}^{s}(h^2)e^{i(s+2p-2\ell)\alpha}\mathfrak{Z}_{\nu+2\ell-s}^{(j)}(k\rho)e^{i(\nu+2\ell-s)\psi} \ ,$$

at which one has to require, corresponding to (9) , that

$$|\rho|e^{\mp \mathrm{Im}\,\psi} > |c_o| + |c|e^{\mp \mathrm{Im}\,\alpha} \ .$$

We also consider the following degeneracies of the addition theorems in Theorem 2 and Theorem 3 :

<u>First case :</u> Let

$$\alpha = 0,\ c \to 0,\ 2h\ \mathrm{Cos}\,z \to kR,\ t \to \phi\ ,\ c\ \mathrm{Cos}(z \pm it) = Re^{\pm i\phi} \ .$$

For the holomorphic solutions (z_o,t_o) and $(\hat{z}_o,\hat{t}_o)$ of

$$c_o\ \mathrm{Cos}(z_o \pm it_o) = Re^{\pm i\phi} + e^{\pm i\beta}\gamma\ \mathrm{Cos}(\zeta \pm i\tau)$$

in

$$\mathcal{L}_e = \left\{(R,\phi)\ \middle|\ \begin{matrix}|R|\ e^{-\mathrm{Im}\,\phi} > \max|e^{i\beta}\gamma\ \mathrm{Cos}(\zeta + i\tau) \pm c_o| \\ |R|\ e^{\mathrm{Im}\,\phi} > \max|e^{-i\beta}\gamma\ \mathrm{Cos}(\zeta - i\tau) \pm c_o|\end{matrix}\right\}$$

and in

$$\mathcal{L}_i = \left\{(R,\phi)\ \middle|\ \begin{matrix}|R|\ e^{-\mathrm{Im}\,\phi} < \min|e^{i\beta}\gamma\ \mathrm{Cos}(\zeta + i\tau) \pm c_o| \\ |R|\ e^{\mathrm{Im}\,\phi} < \min|e^{-i\beta}\gamma\ \mathrm{Cos}(\zeta - i\tau) \pm c_o|\end{matrix}\right\} \ .$$

respectively, we have the following theorems:

If $\hat{h}^2$ is normal value to 0 and 1 , then

(13) $$M_\nu^{(j)}(z_o;h_o)me_\nu(t_o;h_o^2) = \sum_{n=-\infty}^{\infty} D_n\, \mathcal{Z}_{\nu+n}^{(j)}(kR)e^{i(\nu+n)\phi}$$

with

$$D_n = \sum_{m=-\infty}^{\infty} \frac{i^{m+n}}{2\pi}\int_0^{2\pi} me_m(-t+\beta;\hat{h}^2)me_\nu(t;h_o^2)e^{-i(\nu+n)t}dt \cdot M_m^{(1)}(\zeta;\hat{h})me_m(\tau;\hat{h}^2) =$$

$$= \sum_{\ell=-\infty}^{\infty} (-1)^\ell \left\{\sum_{p=-\infty}^{\infty} c_{2p-2\ell}^{2\ell-n}(\hat{h}^2)e^{i(2p-n)\beta}\, c_{2p}^{\nu}(h_o^2)\right\} M_{2\ell-n}^{(1)}(\zeta;\hat{h})me_{2\ell-n}(\tau;\hat{h}^2) .$$

If, respectively, $\hat{h}^2$ is normal value to ν and $\nu+1$, if $\mathrm{Re}(\zeta \pm i\tau) > 0$, and if

(14) $$\min_{\varphi\in[0,2\pi]} \left\{\left|e^{i\beta_\gamma}\operatorname{Cos}(\zeta \pm i\tau + i\varphi)\right|\right\} > |c_o| \qquad \text{(see (9))} ,$$

then

$$M_\nu^{(j)}(\hat{z}_o;h_o)me_\nu(\hat{t}_o;h_o^2) = \sum_{n=-\infty}^{\infty} \hat{D}_n^{(j)}\, J_n(kR)e^{in\phi}$$

with

$$\hat{D}_n^{(j)} = \sum_{m=-\infty}^{\infty} \frac{i^{m+n}}{2\pi}\int_0^{2\pi} me_{\nu+m}(-t+\beta;\hat{h}^2)me_\nu(t;h_o^2)e^{-int}dt\; M_{\nu+m}^{(j)}(\zeta;\hat{h}) \cdot me_{\nu+m}(\tau;\hat{h}^2) .$$

Here (13) and (14) turn out to be rearrangements of (12) and of <u>2.54.</u>, Theorem 2 with a suitable change of the parameter notation.

<u>Second case :</u> Let the (z_o,t_o) - coordinates degenerate into polar coordinates

$$c_o \to 0,\ 2h_o \operatorname{Cos} z_o \to kR_o,\ t_o \to \phi_o$$

$$R_o\, e^{\pm i\phi_o} = c_o \operatorname{Cos}(z_o \pm it_o)$$

with $\alpha = 0$ which implies no restriction. Then, with the solution (R_o,ϕ_o) of

$$R_o e^{\pm i\phi_o} = c \operatorname{Cos}(z \pm it) + e^{\pm i\beta_\gamma}\operatorname{Cos}(\zeta \pm i\tau)$$

in

$$\mathcal{B}_e = \{(z,t)\,|\,\mathrm{Re}\, z \mp \mathrm{Im}\, t > A^\pm\} \qquad \text{(see (2))} ,$$

there holds for normal $\hat{h}^2$ to 0 and 1, and for normal h^2 to ν and $\nu+1$

$$\mathcal{Z}_\nu^{(j)}(kR_o)e^{i\nu\phi_o} = \sum_{n=-\infty}^{\infty}\left\{\sum_{m=-\infty}^{\infty} K_{mn}M_m^{(1)}(\zeta;\hat{h})me_m(\tau;\hat{h}^2)\right\} M_{\nu+n}^{(j)}(z;h)me_{\nu+n}(t;h^2)$$

with

$$K_{mn}(0;h^2;\beta;\hat{h}^2) = \frac{i^{n+m}}{2\pi}\int_0^{2\pi} me_m(-t+\beta;\hat{h}^2)e^{i\nu t}me_{\nu+n}(-t;h^2)dt =$$

$$= \begin{cases} (-1)^\ell \displaystyle\sum_{p=-\infty}^{\infty} c_{2p-2\ell}^{2\ell-n}(\hat{h}^2)e^{i(2p-n)\beta}\, c_{-2p}^{\nu+n}(\hat{h}^2) & (m+n=2\ell) , \\ 0 & (m+n \text{ odd}) . \end{cases}$$

For integer ν and $A^+ \neq A^-$, the terms in the series with the subscripts $\nu + n$ and $-\nu - n$ must be combined. With the presupposition (see 9)) $\mathrm{Re}(\zeta \pm i\tau) > 0$ and with

$$\min_{\varphi \in [0,2\pi]} \left\{ \min_{\sigma \in [-1,1]} \left| e^{\pm i\beta} \gamma \operatorname{Cos}(\zeta \pm i\tau + i\varphi) - c\,\sigma \right| \right\} > 0$$

one obtains for normal $\hat{h}^2$ to ν and $\nu + 1$ and for normal h^2 to 0 and 1, by rearrangement and change of notation, in

$$\mathcal{B}_i = \{(z,t) \mid |\mathrm{Re}\, z \mp \mathrm{Im}\, t| < A^{\pm} = B^{\pm}\} \qquad \text{(see (3))}$$

$$\mathcal{Z}_\nu^{(j)}(kR_o)e^{i\nu\phi_o} = \left\{ \sum_{m=-\infty}^{\infty} L_{mo} M_{\nu+m}^{(j)}(\zeta;\hat{h}) me_{\nu+m}(\tau;\hat{h}^2) \right\} M_o^{(1)}(z;h) me_o(t;h^2) +$$

$$+ \sum_{n=1}^{\infty} \left[\left\{ \sum_{m=-\infty}^{\infty} L_{mn} M_{\nu+m}^{(j)}(\zeta;\hat{h}) me_{\nu+m}(\tau;\hat{h}^2) \right\} M_n^{(1)}(z;h) me_n(t;h^2) + \right.$$

$$\left. + \left\{ \sum_{m=-\infty}^{\infty} L_{m,-n} M_{\nu+m}^{(j)}(\zeta;\hat{h}) me_{\nu+m}(\tau;\hat{h}^2) \right\} M_{-n}^{(1)}(z;h) me_{-n}(t;h^2) \right]$$

with

$$L_{m,n} = \frac{i^{m+n}}{2\pi} \int_0^{2\pi} me_{\nu+m}(-t+\beta;\hat{h}^2) e^{i\nu t} me_n(-t;h^2)\,dt \ .$$

<u>Third case :</u> Let both coordinate systems be degenerate such that the transformation equation becomes with $\alpha = \beta = 0$

$$R_o e^{\pm i\phi_o} = Re^{\pm i\phi} + \gamma \operatorname{Cos}(\gamma \pm i\tau) \ .$$

It is solved in

$$\mathcal{B}_e = \left\{ (R,\phi) \mid |R|\, e^{\pm \mathrm{Im}\,\phi} > |\gamma \operatorname{Cos}(\zeta \pm i\tau)| \right\}$$

by

$$R_o(R,\phi) = R\left(1 + \frac{e^{i\phi}}{R}\gamma \operatorname{Cos}(\zeta - i\tau)\right)^{1/2} \left(1 + \frac{e^{-i\phi}}{R}\gamma \operatorname{Cos}(\zeta + i\tau)\right)^{1/2}$$

and

$$\phi_o(R,\phi) = \phi + \frac{1}{2i} \log \frac{R + e^{-i\phi}\gamma \operatorname{Cos}(\zeta + i\tau)}{R + e^{i\phi}\gamma \operatorname{Cos}(\zeta - i\tau)}$$

(in $(\ )^{1/2}$ and log the principal values are agreed upon). Then for normal values $\hat{h}^2$ to 0 and 1, there results the expansion

$$\mathcal{Z}_\nu^{(j)}(kR_o)e^{i\nu\phi_o} = \sum_{n=-\infty}^{\infty} \left\{ \sum_{\ell=-\infty}^{\infty} (-1)^\ell c_{-2\ell}^{2\ell-n}(\hat{h}^2) M_{2\ell-n}^{(1)}(\zeta;\hat{h}) me_{2\ell-n}(\tau;\hat{h}^2) \right\} \mathcal{Z}_{\nu+n}^{(j)}(kR)e^{i(\nu+n)\phi} \ .$$

2.3. On the computation of the characteristic exponent.

F.W.Schäfke (1961), F.W.Schäfke, R.Ebert and H.Groh (1962) and F.W.Schäfke and D.Schmidt (1966) have developed procedures for the direct computation of the characteristic exponent from the three term recurrence relations. Here we shall outline some of the considerations and results.

The three term recurrence relations for the Fourier coefficients of the even π - periodic Mathieu functions

$$(1)\quad \begin{cases} -2h^2c_1 + \lambda c_0 = 0 \\ -h^2 c_{n+1} + [\lambda-(2n)^2]c_n - h^2 c_{n-1} = 0 & (n = 1,2,3,\ldots) \end{cases}$$

changes with

$$(2)\quad \begin{cases} c_n = (-1)^{n-1}h^{-2n}2^{2n-2}((n-1)!)^2 z_n & (n = 1,2,3,\ldots) \\ c_0 = 2 \end{cases}$$

into

$$(3)\quad \begin{cases} z_1 = \lambda,\ z_2 = \lambda(1-\frac{\lambda}{4}) + \frac{1}{2}h^4 , \\ z_{n+1} = \left(1 - \frac{\lambda}{(2n)^2}\right)z_n - \frac{h^4}{16n^2(n-1)^2} z_{n-1} & (n = 2,3,4,\ldots) . \end{cases}$$

Equally one has in the odd π - periodic case

$$(4)\quad \begin{cases} -h^2c_2 + [\lambda - 4]c_1 = 0 , \\ -h^2c_{n+1} + [\lambda - (2n)^2]c_n - h^2 c_{n-1} = 0 & (n = 2,3,4,\ldots) . \end{cases}$$

This transforms with

$$(5)\quad \begin{cases} c_n = (-1)^{n-1}h^{-2n+2}2^{2n-2}((n-1)!)^2 y_n & (n = 1,2,3,\ldots) \\ c_1 = y_1 = 1 \end{cases}$$

into

$$(6)\quad \begin{cases} y_1 = 1,\ y_2 = (1-\frac{\lambda}{4}) , \\ y_{n+1} = \left(1 - \frac{\lambda}{(2n)^2}\right) y_n - \frac{h^4}{16n^2(n-1)^2} y_{n-1} & (n = 2,3,4,\ldots) . \end{cases}$$

Now one can prove:

1) For every parameter pair λ,h^2 there exist

$$z_I(\lambda,h^2) = \lim_{n\to\infty} z_n(\lambda,h^2)$$

and

$$z_{IV}(\lambda,h^2) = \lim_{n\to\infty} y_n(\lambda,h^2) .$$

These are entire functions which are at most of a normal type of the order $\frac{1}{2}$.

2) One has $z_I(\lambda,h^2) = 0$ if and only if to (λ,h^2) there exists a non-trivial even π - periodic solution; an analogous result holds for $z_{IV}(\lambda,h^2)$ and odd π - periodic solutions.

From this one can obtain

(7) $$z_I(\lambda,h^2) = -\frac{2}{\pi} y_I'\left(\frac{\pi}{2};\lambda,h^2\right)$$

and

(8) $$z_{IV}(\lambda,h^2) = \frac{2}{\pi} y_{II}\left(\frac{\pi}{2};\lambda,h^2\right) ,$$

and because of

$$y_I(\pi) - 1 = 2y_I'(\tfrac{\pi}{2})\, y_{II}(\tfrac{\pi}{2}) ,$$

$$\cos \pi\nu = y_I(\pi) ,$$

where ν is characteristic exponent according to MS 2.13. , there results

(9) $$\sin^2 \nu\frac{\pi}{2} = \frac{\pi^2}{4} z_I(\lambda,h^2)\, z_{IV}(\lambda,h^2) .$$

The convergence can be improved by means of appropriate infinite products. We quote the following two methods:

First method:

$$\gamma_0 = 2,\ \gamma_1 = 1 ,$$

$$\gamma_{n+1} = \gamma_n - \frac{h^4}{[(2n)^2-\lambda][(2n-2)^2-\lambda]}\gamma_{n-1} \qquad (n = 1,2,3,\ldots) ,$$

$$\gamma = \lim \gamma_n ;$$

$$\beta_1 = 1,\ \beta_2 = 1 ,$$

$$\beta_{n+1} = \beta_n - \frac{h^4}{[(2n)^2-\lambda][(2n-2)^2-\lambda]}\beta_{n-1} \qquad (n = 2,3,4,\ldots)$$

$$\sigma = \lim \frac{\beta_n}{\gamma_n} ;$$

$$\sin^2 \nu\frac{\pi}{2} = \gamma^2\sigma \sin^2 \sqrt{\lambda}\frac{\pi}{2} .$$

One has here

$$\gamma_{n+1} - \gamma_n = \mathcal{O}(n^{-4}) ,$$

$$\frac{\beta_{n+1}}{\gamma_{n+1}} - \frac{\beta_n}{\gamma_n} = \mathcal{O}\left(\frac{h^{4n}}{2^{4n}(n!)^2[(n-1)!]^2}\right) .$$

In the exceptional cases $\gamma = 0$ or $\lambda = (2m)^2$ one must suitably modify.

Second Method:

$$\kappa_n = \frac{h^4}{[(2n)^2-\lambda][(2n-2)^2-\lambda]} \qquad (n = 1,2,3,\ldots) ,$$

$$\alpha_o = 0, \alpha_n = \frac{h^4}{[2n(2n-2)-\lambda]^2} \qquad (n = 1,2,3,\ldots) ,$$

$$\zeta_o = 2,\ \zeta_1 = 1,\ \eta_o = 0,\ \eta_1 = 1$$

$$\zeta_{n+1} = \frac{1}{1-\alpha_n}\zeta_n - \frac{\kappa_n}{(1-\alpha_n)(1-\alpha_{n-1})}\zeta_{n-1} \qquad (n = 1,2,3,\ldots) ,$$

$$\eta_{n+1} = \frac{1}{1-\alpha_n}\eta_n - \frac{\kappa_n}{(1-\alpha_n)(1-\alpha_{n-1})}\eta_{n-1} \qquad (n = 1,2,3,\ldots) ,$$

$$\zeta = \lim \zeta_n ,\ \sigma = \lim \frac{\eta_n}{\zeta_n} ,$$

$$t = \frac{\cos\frac{\pi}{2}\sqrt{\lambda+1+h^2}\ \cos\frac{\pi}{2}\sqrt{\lambda+1-h^2}}{\cos^2\frac{\pi}{2}\sqrt{\lambda+1}} ,$$

$$\sin^2 \nu\frac{\pi}{2} = (t\zeta)^2\sigma \sin^2\frac{\pi}{2}\sqrt{\lambda} .$$

Here one has

$$\zeta_{n+1} - \zeta_n = \mathcal{O}(n^{-8})$$

and again

$$\frac{\eta_{n+1}}{\zeta_{n+1}} - \frac{\eta_n}{\zeta_n} = \mathcal{O}\left(\frac{h^{4n}}{2^{4n}(n!)^2[(n-1)!]^2}\right)$$

with due regard to the corresponding exceptional values.

For the details of the procedure, for error estimates and for an ALGOL - program reference is made to the original papers, which were mentioned above.

For a further improvement of convergence see Wagenführer (1976).

2.4. On the eigenvalues for complex h^2 .

The results of MS 1.5., 1.7., 2.25., 2.332. make it desirable to obtain more detailed quantitative and qualitative knowledge of the Riemann surfaces or of the analytic functions in the large, respectively, which emerge from the function elements about $h^2 = 0$. In particular, it is of interest to determine further branch points beyond those - $h^2 = \pm i\ 1{,}468768$, $\lambda = 2{,}0886$ between a_o and a_2 - which have been given by Bouwkamp (1948). (We note that approximate numerical values are always given with significant digits with no rounding off).

With respect to these questions, F.W.Schäfke, R.Ebert, H.Groh and A.Schönhage have, during the years 1959 - 1962, (unpublished) carried out extensive experimental-mathematical, numerical-function-theoretical investigations by means of the electronic computer ER 56 of the University of Cologne. They leant mainly on the recurrence relations MS 2.13. (3) and their derivatives with respect to the parameters, and on the computation of the zeros and branch points of the approximations $z_n(\lambda,h^2)$ with the help of the Newton method.

In particular, there resulted:

$$h^2 = \pm i \cdot 6{,}9289, \ \lambda = 11{,}1904$$

are the branch points on the circles of convergence of b_2 and b_4 ;

$$h^2 = 1{,}9313 \pm i\ 3{,}2376 \ ,$$
$$\lambda = 6{,}1764 \pm i\ 1{,}2317$$

are the branch points on the circles of convergence of a_1 and a_3 ; the values for b_1 and b_3 are then obtained from the symmetry property $b_{2n+1}(h^2) = a_{2n+1}(-h^2)$.

Further purely imaginary branch points of the a_{2n} are:

2) $h^2 = \pm i \cdot 16{,}4711 \ , \ \lambda = 27{,}3191$

3) $h^2 = \pm i \cdot 47{,}8059 \ , \ \lambda = 80{,}6582$

4) $h^2 = \pm i \cdot 95{,}4752 \ , \ \lambda = 162{,}1070$

2), 3) and 4) are on the circle of convergence of a_6, a_{10}, a_{14}, respectively and furnish - in radial direction towards $h^2 = 0$ - a branching with a_4, a_8, a_{12}, respectively. However, for instance for the radius of convergence of a_4, the pair of branch points

$$h^2 = 5{,}1742 \pm i\ 5{,}1042, \ \lambda = 12{,}7997 \pm i\ 2{,}7630$$

is decisive. It furnishes - in radial direction towards $h^2 = 0$ - a branching with a_2 .

Further purely imaginary branch points of the b_{2n+2} are

2) $h^2 = \pm i\ 30{,}0967 \ , \ \lambda = 50{,}4750$,

3) $h^2 = \pm i\ 69{,}5987 \ , \ \lambda = 117{,}8689$,

4) $h^2 = \pm i\ 125{,}4354 \ , \ \lambda = 213{,}3725$.

2), 3) and 4) are on the circle of convergence of b_8, b_{12}, b_{16}, respectively, and furnish - in radial direction towards $h^2 = 0$ - a branching with b_6, b_{10}, b_{14}, respectively. However, for instance for the radius of convergence of b_6, the pair of branch points

$$h^2 = 8{,}1517 \pm i\ 14{,}6932 \ , \ \lambda = 28{,}8886 \pm i\ 4{,}1946$$

is decisive. It furnishes - in radial direction towards $h^2 = 0$ - a branching with b_4 .

We renounce here the reproduction of further numerical data. But we remark that the existence of corresponding further purely imaginary branch points between the a_{4n} and a_{4n+2} (n=0,1,2,...) and analogously between the b_{4n+2} and b_{4n+4} (n=0,1,2,...) can be proved without difficulty.

We note here only the radii of convergence of the first few periodic eigenvalues. They follow from these computations along with error estimates according to the principle of the argument, applied to appropriate discriminants.

The radii of convergence $\rho_n^{(1)}$ of the $a_{2n}(h^2)$ are

n	$\rho_n^{(1)}$
0 or 1	1,4687686137851419923O7
2	7,26814689351672664 4
3	16,471165892263656
4	30,427382096004179
5	47,80596570259757
6	69,9293051764220
7	95,4752727072182
8	125,7662789677
9	159,47921266935

The radii of convergence $\rho_n^{(2)}$ of the $b_{2n+2}(h^2)$ are

n	$\rho_n^{(2)}$
0 or 1	6,9289547587601814796
2	16,80308982535412
3	30,096772837587554
4	48,136381859272
5	69,5987932768953
6	95,8059567052
7	125,435411314
8	159,810254642
9	197,606678692

The radii of convergence $\rho_n^{(3)}$ of the $a_{2n+1}(h^2)$, $b_{2n+1}(h^2)$ are

n	$\rho_n^{(3)}$
0 or 1	3,769957494010357675
2	11,2709852655253850
3	22,85524712159409
4	38,5229250098912
5	58,274138447191
6	82,10894360669
7	110,027369210
8	142,029431279
9	178,11513940

These values are again given with significant digits. See F.W.Schäfke and H.Groh (1962).

It is surprising to see that the radii of convergence seem to increase with the square of the eigenvalue number. Accordingly, the lower bounds of the perturbation theory, for instance

$$\rho_n^{(1)} \geq 2n - 1 \qquad (n = 1,2,3,\ldots) ,$$

were very poor, while the upper bounds, for instance

$$\rho_n^{(1)} \leq 4 \cdot \sqrt{2} \cdot ((2n)^2 - 1) \qquad (n = 2,3,\ldots) ,$$

according to MS 1.5., Theorem 8, $k = 2$ would yield at least the correct order of magnitude.

By evaluating first and second order differences for even n or for odd n in the preceding tables, one is led to the following conjecture for the asymptotic behavior at $n \to \infty$ (with approximate values for the coefficients)

$$\rho_n^{(1)} = 2{,}042(n-1)^2 + 3{,}41(n-1) + 1{,}47 + \mathcal{O}(\tfrac{1}{n}) \qquad (n \text{ odd}),$$

$$\rho_n^{(1)} = 2{,}042(n-1)^2 + 3{,}41(n-1) + 1{,}80 + \mathcal{O}(\tfrac{1}{n}) \qquad (n \text{ even}),$$

$$\rho_n^{(2)} = 2{,}042n^2 + 3{,}41n + 1{,}47 + \mathcal{O}(\tfrac{1}{n}) \qquad (n \text{ odd}),$$

$$\rho_n^{(2)} = 2{,}042n^2 + 3{,}41n + 1{,}80 + \mathcal{O}(\tfrac{1}{n}) \qquad (n \text{ even}),$$

$$\rho_n^{(3)} = 2{,}042(n-\tfrac{1}{2})^2 + 3{,}41(n-\tfrac{1}{2}) + 1{,}55 + \mathcal{O}(\tfrac{1}{n}) .$$

We note that $\pi^{-1/4}e = 2{,}0418\ldots$. This might be a conjecture for an expression of the first coefficient in the above formulas. The second coefficient could be $\frac{5}{3}$ times the first coefficient.

By the way, also the computed branch points exhibit a regular arrangement in sets which have apparently a quadratic law of growth. A relevant theoretical analysis would be very desirable.

Another open problem is the proof that all occurring branchings happen between only two sheets each of the Riemann surface.

From the computational results there also flows the conjecture that all $a_{2n}(h^2)$ $(n = 0,1,2,\ldots)$ belong to one and only one analytic function in the large; and analogously for the other sets of eigenvalues. This can be proved (see below).

We mention that numerical results on branch points and radii of convergence have also been given by G.Blanch and D.S.Clemm (1969). But their data are less extensive.

We give now F.W.Schäfke's proof (1975) of the

Theorem : All $a_{2n}(h^2)$ $(n = 0,1,2,\ldots)$ belong to one analytic function in the large; therefore $y_I'(\frac{\pi}{2};\lambda,h^2)$ is irreducible, in the sense that a decomposition into a product of entire functions, both of which yield zeros of $y_I'(\frac{\pi}{2};\lambda,h^2)$, is impossible.

Analogous statements hold

for the $b_{2n+2}(h^2)$ $(n = 0,1,2,\ldots)$,

for the $a_{2n+1}(h^2)$ $(n = 0,1,2,\ldots)$,

for the $b_{2n+1}(h^2)$ $(n = 0,1,2,\ldots)$, and

for the $\lambda_{\nu+2n}(h^2)$ $(n \in \mathbb{Z})$ with real non-integer ν .

We give the proof for the $a_{2n}(h^2)$; it is completely analogous in the other cases.

Let $\emptyset \neq M \subseteq \mathbb{N}_0 = \{0,1,2,\ldots\}$ such that the $a_{2n}(h^2)$ $(n \in M)$ in arbitrary analytic continuation from $h^2 = 0$ to $h^2 = 0$ produce only such function elements which belong to the same set. For $m \in \mathbb{N}$ we consider

$$f_m(h^2) := \sum_{\substack{n \in M \\ n \leq m}} a_{2n}(h^2) = \sum_{k=0}^{\infty} f_{m,k} h^{4k}$$

For $|h^2| < 2m+1$, all $a_{2n}(h^2)$ with $n \geq m+1$ are holomorphic. Therefore the radius of convergence r_m of the symmetric function f_m obviously satisfies

$$r_m \geq 2m + 1 \ .$$

It is known that

$$|f_{m,2}| \leq \frac{2}{r_m^4} \max_{|z|=r_m} |\mathrm{Im}\, f_m(z)| \ .$$

Therefore one has, because of

$$|\mathrm{Im}\, \lambda| \leq 2|h^2| \ ,$$

for every pair of eigenvalues the estimate

$$|f_{m,2}| \leq \frac{4(m+1)r_m}{r_m^4} \leq \frac{4(m+1)}{(2m+1)^3} \ .$$

This leads evidently to

$$(*) \qquad \sum_{n \in M} \lambda_{2n,4} = 0 \ ,$$

which must hold in particular, for $M = \mathbb{N}_0$. If now one considers the known coefficients of h^8 in the expansions of the $a_{2n}(h^2)$, then one recognizes: one and only one is negative, all others are positive. Consequently there results that $(*)$ is possible only for $M = \mathbb{N}_0$. This is the assertion.

In a similar way, the other cases can be treated. In concluding we note the relations

$$\sum_{n=0}^{\infty} \left(a_{2n}(h^2) - (2n)^2\right) = 0 \ ,$$

$$\sum_{n=0}^{\infty} \left(b_{2n+2}(h^2) - (2n+2)^2\right) = 0 \ ,$$

$$\sum_{n=0}^{\infty} \left(a_{2n+1}(h^2) - (2n+1)^2\right) = h^2 \ ,$$

$$\sum_{n=0}^{\infty} \left(b_{2n+1}(h^2) - (2n+1)^2\right) = -h^2 \ .$$

They are closely connected with the preceding considerations and also easily provable.

2.5. Improved estimates of the radii of convergence.

As has been just described in 2.4., the numerical results show that the radii of convergence of the $a_m(h^2)$, $b_m(h^2)$ seem to grow with m according to a quadratic law. In contrast to this, the known lower estimates obtained from the perturbation theory give a law of growth proportional to the distance of adjacent eigenvalues, that is to m only. The big difference between numerical experience and what has been proved theoretically motivates, of course, intensive attempts in order to gain improved theoretical estimates.

A first result in this direction has been given by A. Schönhage (unpublished). He studied the $b_{2n+2}(h^2)$ and, by proving a lim-inf-statement, he could prove at least the existence of lower bounds proportional to $m \cdot \log m$. We give here a simple evaluation and improvement of this idea.

Let $\nu \in \mathbb{C}$. Then ν is characteristic exponent to (λ, h^2) if and only if

$$(1) \qquad (\lambda-(\nu+2n)^2)\gamma_n = h^2(\gamma_{n+1}+\gamma_{n-1}) \qquad (n \in \mathbb{Z})$$

has a non-trivial solution with

$$(2) \qquad \gamma_n \to 0 \qquad (|n| \to \infty) \ .$$

We introduce the abbreviation

$$(3) \qquad \lambda_n := (\nu+2n)^2 \qquad (n \in \mathbb{Z}) \ .$$

Then one proves for $\lambda \neq \lambda_n$ $(n \in \mathbb{Z})$ by induction with respect to $k \in \mathbb{N}_0$ the following iterated form of (1) :

$$(4) \qquad \gamma_n = \sum_{j=0}^{k} \binom{k}{j} \frac{\gamma_{n-k+2j-1}+\gamma_{n-k+2j+1}}{\prod\limits_{\sigma=0}^{k}(\lambda-\lambda_{n+j-\sigma})} \cdot h^{2k+2} - \sum_{\kappa=2}^{k} \sum_{j=1}^{\kappa-1} \frac{\binom{\kappa-2}{j-1} 4\kappa(\kappa-1)\gamma_{n-\kappa+2j}}{\prod\limits_{\sigma=0}^{\kappa}(\lambda-\lambda_{n+j-\sigma})} \cdot h^{2\kappa} \ .$$

For $k = 0$ this is just (1). For the conclusion $k \longrightarrow k+1$ one introduces (1) appropriately into (4) and proceeds as follows:

$$\sum_{j=0}^{k} \binom{k}{j} \frac{\gamma_{n-k+2j-1}+\gamma_{n-k+2j+1}}{\prod\limits_{\sigma=0}^{k}(\lambda-\lambda_{n+j-\sigma})} =$$

$$= \sum_{j=0}^{k+1} \left[\binom{k}{j} \prod_{\sigma=0}^{k} (\lambda-\lambda_{n+j-\sigma})^{-1} + \binom{k}{j-1} \prod_{\sigma=0}^{k} (\lambda-\lambda_{n+j-\sigma-1})^{-1}\right] \gamma_{n-k+2j-1} =$$

$$= \sum_{j=0}^{k+1} \frac{\binom{k}{j}(\lambda-\lambda_{n+j-k-1}) + \binom{k}{j-1}(\lambda-\lambda_{n+j})}{\prod\limits_{\sigma=0}^{k+1}(\lambda-\lambda_{n+j-\sigma})} \gamma_{n-k+2j-1} =$$

$$= \sum_{j=0}^{k+1} \frac{\binom{k+1}{j}(\lambda-\lambda_{n-k+2j-1}) + S(n,j,k)}{\prod_{\sigma=0}^{k+1} (\lambda-\lambda_{n+j-\sigma})} \Upsilon_{n-k+2j-1} =$$

$$= h^2 \sum_{j=0}^{k+1} \binom{k+1}{j} \frac{\Upsilon_{n-k+2j-2}+\Upsilon_{n-k+2j}}{\prod_{\sigma=0}^{k+1} (\lambda-\lambda_{n+j-\sigma})} +$$

$$+ \sum_{j=0}^{k+1} \frac{S(n,j,k)}{\prod_{\sigma=0}^{k+1} (\lambda-\lambda_{n+j-\sigma})} \Upsilon_{n-k+2j-1} .$$

There

$$S(n,j,k) := \binom{k}{j}(\lambda-\lambda_{n+j-k-1}) + \binom{k}{j-1}(\lambda-\lambda_{n+j}) - \binom{k+1}{j}(\lambda-\lambda_{n-k+2j-1}) =$$

$$= \binom{k+1}{j}(\nu+2(n+2j-k-1))^2 - \binom{k}{j}(\nu+2(n+j-k-1))^2 - \binom{k}{j-1}(\nu+2(n+j))^2 =$$

$$= - 4\binom{k-1}{j-1} k(k+1)$$

and $= 0$ for $j = 0$ or $k = 0$.

This result is obtained straightforwardly by simple algebra for the coefficients of the powers of $(\nu + 2n)$:

$$\binom{k+1}{j} - \binom{k}{j} - \binom{k}{j-1} = 0 ,$$

$$\binom{k+1}{j} (2j-k-1) - \binom{k}{j}(j-k-1) - \binom{k}{j-1}j = \binom{k+1}{j}(j-k-1) + \binom{k}{j}(k+1) = 0 ,$$

$$\binom{k+1}{j}(2j-k-1)^2 - \binom{k}{j}(j-k-1)^2 - \binom{k}{j-1}j^2 =$$

$$= \binom{k+1}{j}\left[2j(j-k-1)+(j-k-1)^2\right] + \binom{k}{j}\left[2j(k+1) - (k+1)^2\right] =$$

$$= \binom{k+1}{j}(k+1-j)^2 - \binom{k}{j}(k+1)^2 = \binom{k-1}{j-1}\left[\frac{k(k+1)}{j}(k+1-j) - \frac{k(k+1)}{j}(k+1)\right] =$$

$$= - \binom{k-1}{j-1} k(k+1) .$$

This completes the proof of (4).

It is an essential feature of (4) that only products of the $\lambda - \lambda_n$ with adjoining different subscripts occur and that in the second term the number of factors in the denominator exceeds by 1 the respective exponent of h^2. This ensures the success of the following estimates.

We assume in the following

(5) $$0 < \delta_m < \min\{|\lambda_n - \lambda_m| : \lambda_n \neq \lambda_m\}$$

and set

(6) $$a_{mn} := \Big| |\lambda_n - \lambda_m| - \delta_m \Big| \qquad (n \in \mathbb{Z},\ m \in \mathbb{Z}) .$$

Now let $m \in \mathbb{Z}$ and

$$|\lambda - \lambda_m| = \delta_m , \quad |h^2| \leq r < \infty ,$$

and let ν be the characteristic exponent to (λ, h^2) and $0 < |\gamma_n| = \max\{|\gamma_k| : k \in \mathbb{Z}\}$, where (γ_k) is a solution of (1) with (2). Then it follows from (4) that

(7) $$1 \leq 2r^{k+1} \sum_{j=0}^{k} \binom{k}{j} \prod_{\sigma=0}^{k} a^{-1}_{m,n+j-\sigma} + \\ + 4 \sum_{\kappa=2}^{k} \sum_{j=1}^{\kappa-1} \binom{\kappa-2}{j-1} \kappa(\kappa-1) \prod_{\sigma=0}^{\kappa} a^{-1}_{m,n+j-\sigma} \cdot r^{\kappa} .$$

Let now, for $m \in \mathbb{Z}$, $n \in \mathbb{Z}$, r_{mn} be the supremum of those $r \geq 0$, for which there exists a $k \in \mathbb{N}_0$ such that (7) does not hold. Then it is obvious that for $|\lambda - \lambda_m| = \delta_m, |h^2| < r_{mn}$, (1) and (2) have no non-trivial solution for which $|\gamma_n|$ has maximal magnitude. Now let

(8) $$0 < r_m \leq r_{nm} \qquad (n \in \mathbb{Z}).$$

Then for

$$|\lambda - \lambda_m| = \delta_m, \ |h^2| < r_m$$

ν cannot be characteristic exponent to (λ, h^2).

But this means that in case of $\nu \notin \mathbb{Z}$ the radius of convergence of $\lambda_{\nu+2m}(h^2)$, in case of $\nu = 0$ the radii of convergence of a_{2m}, b_{2m}, and in case of $\nu = 1$ the radii of convergence of a_{2m+1}, b_{2m+1}, are at least r_m. For: the respective entire characteristic function $\Delta(\lambda, h^2)$ has within

$$|\lambda - \lambda_m| < \delta_m$$

for $h^2 = 0$ precisely one zero and consequently also for $|h^2| < r_m$ precisely one holomorphic zero, which is $\lambda_{\nu+2m}(h^2)$ or $a_{2m}(h^2)$ ..., respectively, because just for $|h^2| < r_m$ we have $\Delta(\lambda, h^2) \neq 0$ on $|\lambda - \lambda_m| = \delta_m$. We note

<u>Theorem 1 :</u> Let $\nu \in \mathbb{C}$ and let r_m for $m \in \mathbb{Z}$ be given according to the mentioned instruction. Then the functions

$\lambda_{\nu+2m}(h^2)$	$(\nu \notin \mathbb{Z},\ m \in \mathbb{Z})$	
$a_{2m}(h^2)$	$(m \in \mathbb{N}_0)$,	$(\nu = 0)$
$b_{2m}(h^2)$	$(m \in \mathbb{N})$,	$(\nu = 0)$
$a_{2m+1}(h^2)$, $b_{2m+1}(h^2)$	$(m \in \mathbb{N}_0)$,	$(\nu = 1)$

are holomorphic for $|h^2| < r_m$ and satisfy the estimate

$$|\lambda_{\nu+2m}(h^2) - (\nu+2m)^2| < \delta_m \quad ,$$

and analogous estimates in the other cases. -

Now there arises the problem to actually specify suitable δ_m, r_m. This is possible by further estimates. We assume

$$|\operatorname{Re} \nu| \leqq 1$$

without loss of generality and restrict δ_m instead of (5), for instance, by

$$(9) \qquad \frac{1}{4} \min_n\{|\lambda_n - \lambda_m| : \neq 0\} \leqq \delta_m \leqq \frac{1}{2} \min\{\ldots\} \quad .$$

In an actual case one may possibly modify this restriction with the goal of obtaining more favorable constants in the estimates.

As we are now going to demonstrate, estimates of the following kind are meaningful and possible in case of (9) :

With

$$(10) \qquad \begin{array}{l} k_0 \in \mathbb{N}_0,\ m_0 \in \mathbb{N},\ \alpha > 0 \ , \\ 0 < \beta \leqq 1,\ 0 < \gamma \leqq 1,\ 0 < \delta \leqq 1 \end{array}$$

there shall hold

$$(11) \qquad k_0 + 1 < \gamma m_0,\ 2\delta m_0 \geqq 1 \ ;$$

and for $m \in \mathbb{Z}$ with $|m| \geqq m_0$ and $n \in \mathbb{Z}$ it is required that

$$(12.1) \qquad a_{mn} \geqq 4\alpha|m|\,(||n| - |m|| + 1) \qquad (k_0 < ||n| - |m|| \leqq \gamma|m|-1),$$

$$(12.2) \qquad a_{mn} \geqq 4\alpha\beta|m|(||n|-|m|| + 1) \qquad (||n|-|m|| \leqq k_0) \ ,$$

$$(12.3) \qquad a_{mn} \geqq 4\alpha|m|(\delta|m| + \tfrac{1}{2}) \qquad (||n|-|m|| + 1 > \gamma|m|) \ .$$

Now it follows from (10),(11), (12) for $|m| \geqq m_0$, $p + 2 \leqq 2\delta|m| + 1$ and all $n \in \mathbb{Z}$, that

$$(13) \qquad \prod_{q=n}^{n+p} a_{mq} \geqq \frac{1}{2}\, \beta^{2k_0+1}\, (2\alpha|m|)^{p+1}(p+2)! \ .$$

For the proof we assume at first that (12.1) holds for $\hat{p}$ factors in the left member of (13). Then one certainly has because of $p + 1 \leqq 2m$

$$\prod_{\substack{q=n \\ (q \text{ with } (12.1))}}^{n+p} a_{mq} \geqq (4\alpha|m|)^{\hat{p}+1}\, p_1!\,p_2! \ .$$

with $p_1 + p_2 = \hat{p} + 2$, and $p_1 = p_2$ or $p_1 = p_2 + 1$; therefore

$$\prod_{q=n}^{n+p} a_{mq} \geq \frac{1}{2}(2\alpha|m|)^{\hat{p}+1}(\hat{p}+2)! \quad .$$

(q with (12.1))

If (12.1) or (12.2) holds for altogether $\hat{p}$ factors, then a factor β must be introduced at most $(2k_o+1)$ - times in the right member of the last inequality. There remain now $p-\hat{p}$ factors for which (12.3) holds. Thus we have finally

$$\prod_{q=n}^{n+p} a_{mq} \geq \frac{1}{2}\beta^{2k_o+1}(2\alpha|m|)^{p+1}(\hat{p}+2)!(2\delta|m|+1)^{p-\hat{p}} \ .$$

Because of $p+2 \leq 2\delta|m|+1$ this yields just (13). With (13) there results from (7) for $k+2 \leq 2\delta|m|+1$

$$\text{(14)} \qquad \beta^{2k_o+1} \leq \frac{2}{(k+2)!}\left(\frac{r}{\alpha|m|}\right)^{k+1} + \frac{1}{\alpha|m|}\sum_{\kappa=2}^{k}\frac{\kappa(\kappa-1)}{(\kappa+2)!}\left(\frac{r}{\alpha|m|}\right)^{\kappa} \quad .$$

If one chooses now

$$2\delta|m| < k+2 \leq 2\delta|m|+1 \quad ,$$

then (14) leads to

$$\text{(15)} \qquad 1 \leq \beta^{-(2k_o+1)}\left[\frac{1}{\delta}+\frac{1}{\alpha}\right]\frac{1}{|m|}\exp\left(\frac{r}{\alpha|m|}\right) \quad .$$

Thus one obtains with the presuppositions (10), (11), (12) :

<u>Theorem 2 :</u> In Theorem 1 one can choose for $|m| \geq m_o$, if > 0 ,

$$r_m := \alpha|m|\log\left[|m|\ \beta^{2k_o+1}\frac{\alpha\delta}{\alpha+\delta}\right] .$$

In particular, one can now start from the fact that, with a constant $\alpha > 0$ for $|m| \geq m_o$ and $n \in \mathbb{Z}$, there holds

$$\text{(16)} \qquad a_{mn} \geq 4\alpha|m|(||n|-|m||+1) \ .$$

In this case one can choose

$$k_o = 0, \ \beta = \gamma = \delta = 1$$

and obtains herewith, if > 0 , the possible choice

$$\text{(17)} \qquad r_m := \alpha|m|\log\left(\frac{\alpha}{\alpha+1}|m|\right) \quad .$$

However, the possible value of α in (17) will obviously be relatively small. But the value of α is essential for the growth of r_m at $m \to \infty$. Therefore it is of interest to make α in (12) as large as possible. We show in this direction

<u>Lemma 3 :</u> Choose γ arbitrarily in $0 < \gamma \leq 1$. Then there exist

$$m_{o1} \in \mathbb{N} \ , \ c_o > 0, \ c_1 > 0$$

with

$$c_1|m| \leq \delta_m \leq 4c_o(2-\gamma)|m| \qquad (|m| \geq m_{o1}) \ .$$

Then one determine $\mathbb{N} \ni m_{o2} \geqq m_{o1}$, $k_o \in \mathbb{N}$ with

$$c_o < k_o < \gamma m_{o2} - 1 \quad ,$$
$$c_o(2-\gamma) < \frac{1}{2}\gamma m_{o2} - 2 \quad .$$

If now one sets

$$\alpha := (2-\gamma)\,\frac{k_o - c_o}{k_o + 1} \quad ,$$
$$\delta := \min\left(\frac{1}{2}\,\frac{\gamma}{\alpha}\,,\ 1\right) \quad ,$$
$$\beta := \frac{c_1}{4\alpha(k_o + 1)}$$

and chooses $\mathbb{N} \ni m_o \geqq m_{o2}$ with

$$2\delta m_o \geqq 1 \quad ,$$

then with these constants (10), (11), (12) are satisfied.

Proof :

(i) $k_o < ||n| - |m|| \leqq \gamma|m| - 1$:

Then

$$a_{mn} \geqq 4|n-m||n+m+\mathrm{Re}\ \nu| - 4c_o(2-\gamma)|m|$$
$$\geqq 4[|n-m|(|n+m|-1) - c_o(2-\gamma)|m|] \quad .$$

Now one has

$$|n| \geqq |m|(1-\gamma) + 1$$

and either $|n-m| = ||n|-|m||$ and

$$|n-m| \geqq k_o + 1,\ |n+m| > |m|(2-\gamma) + 1 \ ,$$

or $|n+m| = ||n|-|m||$ and

$$|n-m| \geqq |m|(2-\gamma) + 1,\ |n+m| \geqq k_o + 1 \quad .$$

In both cases there follows

$$a_{mn} \geqq 4(2-\gamma)|m|(||n|-|m|| - c_o - 1)$$
$$\geqq 4(2-\gamma)|m|\,\frac{k_o - c_o}{k_o + 1}(||n|-|m|| + 1) \ .$$

With the α introduced in Lemma 3 this agrees with (12.1).

(ii) $||n|-|m|| \leqq k_o$:

One has in any case

$$a_{mn} \geqq \delta_m \geqq c_1|m| \qquad (|m| \geqq m_o) \quad ,$$

therefore with β from Lemma 3 the property (12.2) follows.

(iii) $||n|-|m| + 1 > \gamma|m|$:

One has either

$$|n-m| \geqq \gamma|m| - 1,\ |n+m| \geqq |m|$$

or

$$|n-m| \geqq |m|, \quad |n+m| \geqq \gamma|m| - 1 \quad .$$

But

$$(\gamma|m|-1)(|m|-1) > |m|(\gamma|m|-2) \quad .$$

Therefore

$$a_{mn} \geqq 4[|n-m||n+m + \operatorname{Re} \nu| - c_o(2-\gamma)|m|]$$
$$\geqq 4[(\gamma|m|-2)|m| - c_o(2-\gamma)|m|] \quad .$$

Now it was assumed that

$$c_o(2-\gamma) \leqq \frac{1}{2}\gamma\,|m| - 2 \quad .$$

Therefore there results

$$a_{mn} \geqq 2\gamma|m|^2$$

with a right member which, due to the choice of δ in Lemma 3, is larger than the right member of (12.3).

In any case one can choose γ arbitrarily small and k_o arbitrarily large. Therefore α can approach 2 from below as closely as one desires. Thus we have

Theorem 3 : With the assumption (9) one can in Theorem 1 give values r_m such that

$$\lim\inf \frac{r_m}{|m|\log|m|} \geqq 2 \quad .$$

It is noteworthy that this statement holds for all $\nu \in \mathbb{C}$.

We give finally the result for a special case: If $\nu = 0$ and $m \geqq 2$, one can choose $\delta_m := 2m$. Then (16) holds with $\alpha = \frac{1}{2}$ and (17) yields

$$r_m := \frac{m}{2} \log \frac{m}{3} \qquad (m \geqq 4) \quad .$$

In concluding we remark emphatically that most of the estimates were pretty rough, with the goal to arrive at simple formulas in Theorem 2 and in (17). Of course, one can proceed in a much more refined way, and this already with (5), (6), (7). But apparently no qualitative improvement can be attained; in particular, it seems that the statement of Theorem 3 cannot be improved in this way.

2.6. Asymptotic estimates for large h^2.

The known asymptotic formulas and series for eigenvalues and eigenfunctions for large real h^2 are throughout statements of approximate values without error estimates. At best, they are a sequence of limit-statements. As a rule, their numerical usefulness is demonstrated only in an exemplary manner by comparing them with exact values which are determined in a different way, in the expectation that the errors for other large values of h^2 will be correspondingly small.

This is of course extremely unsatisfactory. But this is for the simple reason that already the analytical effort in the derivation of formulas is pretty large and requires in its performance an oversize amount of tenacity, hard work and skill

M.Kurz has for the first time carried through error estimates in an complete and quite satisfactory manner. Naturally, not in the least one could think of deriving optimal bounds. Rather aspects of a practicable execution and application had to come to the fore. We give in the following only the achieved results. Thus an essential point in the work of M.Kurz is completely lost - an account of the difficulties of the undertaking to attack in this field the problem of the error estimates.

We employ in the following

$$(1)\qquad \tilde{D}_p(\zeta) := \begin{cases} 0 & (p \in -\mathbb{N}) , \\ 2^{p/2} D_p(\zeta) = \left(-2^{1/2}\right)^p e^{\frac{\zeta^2}{4}} \dfrac{d^p}{d\zeta^p} e^{-\frac{\zeta^2}{2}} & (p \in \mathbb{N}_0) . \end{cases}$$

Then one sets for $\ell \in \mathbb{N}_0$, $n \in \mathbb{N}_0$

$$(2)\qquad \lambda_{n,\ell}(h) := -2h^2 + (4n+2)h + \sum_{r=0}^{\ell} \gamma_{n,r} h^{-r} ,$$

$$(3)\qquad y_{n,\ell}(z;h) := \sum_{r=0}^{\ell} h^{-r} \sum_{\nu=-2r}^{2r} g_{n,r,\nu}\, \tilde{D}_{n+2\nu}(\zeta) = \sum_{\nu=-2\ell}^{2\ell} \alpha_{n,\ell,\nu}(h) \tilde{D}_{n+2\nu}(\zeta)$$

with

$$(4)\qquad \zeta = 2h^{1/2} \cos z$$

and

$$(5)\qquad \alpha_{n,\ell,\nu} = \sum_{2\ell \geqq 2r \geqq |\nu|} h^{-r} g_{n,r,\nu}$$

and determines the coefficients from

$$(6)\qquad \begin{cases} g_{n,o,o} = 1 , \\ g_{n,o,\nu} = 0 \quad (\nu \neq 0),\ g_{n,r,o} = 0 & (r > 0) , \end{cases}$$

$$(7)\qquad \begin{cases} (-8\nu) g_{n,r+1,\nu} = - \sum_{s=0}^{r} g_{n,s,\nu}\, \gamma_{n,r-s} + \frac{1}{16} g_{n,r,\nu-2} + \\ \quad + \frac{1}{4} g_{n,r,\nu-1} - \frac{1}{2}\left[(n+2\nu+\frac{1}{2})^2 + \frac{1}{4}\right] g_{n,r,\nu} - \\ \quad - (n+2\nu+2)(n+2\nu+1) g_{n,r,\nu+1} + (n+2\nu+4)(n+2\nu+3)(n+2\nu+2)(n+2\nu+1) g_{n,r,\nu+2} , \end{cases}$$

$$(8)\qquad \begin{cases} \gamma_{n,r} = \frac{1}{16} g_{n,r,-2} + \frac{1}{4} g_{n,r,-1} - \frac{1}{2}\left[(n+\frac{1}{2})^2 + \frac{1}{4}\right] g_{n,r,o} - \\ \quad - (n+2)(n+1) g_{n,r,1} + (n+4)(n+3)(n+2)(n+1) g_{n,r,2} . \end{cases}$$

Then with

$$(9)\qquad L_\lambda(h) f(z) := f''(z) + (\lambda - 2h^2 \cos 2z) f(z)$$

there results

$$(10)\qquad L_{\lambda_{n,\ell}(h)}(h)y_{n,\ell}(z;h) = h^{-\ell}\sum_{\substack{\nu=-2\ell-2\\ \nu\neq 0}}^{2\ell+2} \beta_{n,\ell,\nu}(h)\widetilde{D}_{n+2\nu}(\zeta)$$

with

$$(11)\qquad \beta_{n,\ell,\nu}(h) = 8\nu g_{n,\ell+1,\nu} + \sum_{2\ell\geqq 2r\geqq|\nu|}\ \sum_{s=r-\ell+1}^{\ell} h^{\ell-r-s} g_{n,r,\nu}\,\gamma_{n,s}$$

Now one can show:

If one sets

$$(12)\qquad F(h;n,\ell) := 9\cdot 2^{-3\ell-2}(\ell+1)!\left(\frac{(n+4\ell+4)!}{n!}\right)^{1/2} h^{-\ell} ,$$

with the restrictions

$$(13)\qquad h \geqq \frac{1}{3}(n+4\ell+4)^2\ell$$

and

$$(14)\qquad F(h;n,\ell) < \frac{3}{2}$$

then there holds the estimate

$$(5)\qquad |\lambda_{n,\ell}(h) - a_n(h^2)| \leqq F(h;n,\ell) .$$

If one sets

$$\langle f,g\rangle = \int_0^{\pi/2} f(x)g(x)\,dx ,$$

$$(16)\qquad \|f\|_2 = \langle f,f\rangle^{1/2}$$

$$\|f\|_\infty = \max\left\{|f(x)| : 0 \leqq x \leqq \frac{\pi}{2}\right\}$$

and determines $\sigma_{n,\ell}(h)$ from

$$(17)\qquad \begin{cases} \|\sigma_{n,\ell}(h)y_{n,\ell}(h)\|_2 = \|ce_n(\cdot;h^2)\|_2 , \\ \langle\sigma_{n,\ell}(h)y_{n,\ell}(h),\ ce_n(\cdot;h^2)\rangle > 0 , \end{cases}$$

then one can show for

$$(18)\qquad h \geqq \frac{1}{3}(n+4\ell+6)^2\ell$$

and with the assumption (14) that

$$(19)\qquad \|\sigma_{n,\ell}(h)y_{n,\ell}(h) - ce_n(\cdot;h^2)\|_2 < \left(\frac{\pi}{98}\right)^{1/2} h^{-1}F(h;n,\ell) .$$

Finallly with

$$(20)\qquad \begin{cases} h \geqq \frac{1}{3}(n+4\ell+14)^2(\ell+2) , \\ F(h;n+2,\ell+2) < \frac{3}{2} , \\ F(h,n+2,\ell) < \frac{3}{2} \end{cases}$$

there holds the estimate

(21) $$\|\sigma_{n,\ell}(h)y_{n,\ell}(h) - ce(\cdot;h^2)\|_\infty < \left(\frac{h}{2}\right)^{5/4} F(h;n,\ell+2)\{2^{n+4\ell}+2\} \quad .$$

Starting point for these estimates is the proof that the ratio of the norms $\| \|_2$ of (10) and of (3) is not greater than (12).

We note in passing that many formal results on Mathieu functions for large h^2 can be found in various papers by R.B.Dingle and H.J.W.Müller (1962), (1964) and by H.J.W.Müller (1962), (1964). Among others, also G.Blanch (1960), R.Sips (1965) and A.Sharples (1967), (1970) should be mentioned.

2.7. On the power series of the eigenvalues.

In MS the power series of the eigenvalues $\lambda_{\nu+2n}(h^2)$ and those of the $a_m(h^2)$, $b_m(h^2)$ in MS 2.25. (35) and (36), (37) stand in juxtaposition without an apparent link. The only common point seems to be that all these power series are derived by similar methods from various three term recurrence relations. One can, however, easily realize that the series expansions for integer ν can be obtained from those for non-integer ν, at least in principle.

One starts with the entire function of λ, h^4, ν^2

$$\Delta(\lambda,h^4,\nu^2) := y_I(\pi;\lambda,h^2) - \cos \pi\nu \quad .$$

It has for $h^4 = \nu^2 = 0$ the double zero $\lambda = (2n)^2$ with $n \in \mathbb{N}$ and differs from zero for instance on the circle $|\lambda - (2n)^2| = 2n$, and this holds also for small h^4 and ν^2. Therefore the functions

$$\frac{1}{2\pi i} \oint_{|\lambda-(2n)^2|=2n} \lambda^\kappa \frac{\Delta_\lambda(\lambda,h^4,\nu^2)}{\Delta(\lambda,h^4,\nu^2)} d\lambda \qquad (\kappa=0,1,2)$$

are, for (h^4,ν^2) in a neighborhood of $(0,0)$ holomorphic functions of the variables. Their values are for small $\nu \neq 0$

$$\lambda^\kappa_{\nu+2n}(h^2) + \lambda^\kappa_{-\nu+2n}(h^2) \qquad (\kappa=1,2)$$

and for $\nu = 0$

$$a^\kappa_{2n}(h^2) + b^\kappa_{2n}(h^2) \qquad (\kappa=1,2) \quad .$$

Therefore similar statements hold for the symmetrical polynomials of these zeros of $\Delta(\cdot,h^4,\nu^2)$, so for

$$d^2_{2n}(h^4,\nu^2) := \begin{cases} \left(\lambda_{\nu+2n}(h^2) - \lambda_{-\nu+2n}(h^2)\right)^2 & (\nu \neq 0) \quad , \\ \left(a_{2n}(h^2) - b_{2n}(h^2)\right)^2 & (\nu = 0) \quad . \end{cases}$$

Therefore, for small $\nu \neq 0, h^2$, the sum and the square of the difference of the functions $\lambda_{\pm\nu+2n}(h^2)$ give holomorphic functions of these variables for (h^4,ν^2) near $(0,0)$, which for $\nu = 0$ turn into the sum and into the square of the difference of $a_{2n}(h^2)$, $b_{2n}(h^2)$. Hence one can simply set $\nu = 0$ in the respective coefficients of the series expansions in powers of h^4. An analogous result holds in the vicinity of $(2n+1)^2$ $(n \in \mathbb{N}_0)$. For $a_0(h^2)$ one can obviously directly set $\nu = 0$ in $\lambda_\nu(h^2)$. Thus one can evaluate the power series for $a_m(h^2)$, $b_m(h^2)$ from MS 2.25., (35).

We give the beginning terms of the mentioned expansions as they are obtained from MS 2.25. (35) by J.Meixner (unpublished):

$$\frac{1}{2}\left[\lambda_{-\nu+n}(h^2) + \lambda_{\nu+n}(h^2)\right] = n^2 + \nu^2 + \frac{h^4}{2}\;\frac{n^2 - 1 + \nu^2}{[(n+1)^2-\nu^2][(n-1)^2-\nu^2]} +$$

$$+ \frac{h^8}{32\;[(n+1)^2-\nu^2]^3[(n-1)^2-\nu^2]^3[(n+2)^2-\nu^2][(n-2)^2-\nu^2]} \times$$

$$\times \;\{(n^2-1)^3(n^2-4)(7+5n^2) + (n^2-1)(71-689n^2+121n^4+65n^6)\nu^2 -$$

$$- 10(n^2-1)(4-77n^2+7n^4)\nu^4 + (26+56n^2-70n^4)\nu^6$$

$$+ (65n^2-28)\nu^8 + 5\nu^{10}\} + \mathcal{O}(h^{12}) \quad ,$$

$$\frac{1}{4}\left[\lambda_{-\nu+n}(h^2) - \lambda_{\nu+n}(h^2)\right]^2 = 4n^2\nu^2 - \frac{4n^2\nu^2h^4}{[(n+1)^2-\nu^2][(n-1)^2-\nu^2]} +$$

$$+ \frac{n^2\nu^2h^8}{4[(n+1)^2-\nu^2]^3[(n-1)^2-\nu^2]^3[(n+2)^2-\nu^2][(n-2)^2-\nu^2]} \times$$

$$\times \quad \{(n^2-1)^2(-11n^4-20n^2+175) + (n^2-1)(-36n^4-262n^2+370)\nu^2 +$$

$$+ (94n^4-226n^2+204)\nu^4 - (36n^2-2)\nu^6 - 11\nu^8\} + \mathcal{O}(h^{12}) \;.$$

This method can also be applied to the appertaining functions. We follow the above consideration about $((2n)^2,0,0)$: Then one can consider, say, the entire function of z,λ,h^2,ν - se MS 2.13. Theorem 1 -

$$f(z,\lambda,h^2,\nu) \;:=\; y_I(z+\pi;\lambda,h^2) - e^{-\pi i\nu}y_I(z;\lambda,h^2)$$

and with it

$$\frac{1}{2\pi i}\oint_{|\lambda-(2n)^2|=2n} (f(z,\lambda,h^2,\nu))^\kappa\;\frac{\Delta_\lambda(\lambda,h^4,\nu^2)}{\Delta(\lambda,h^4,\nu^2)}\,d\lambda \qquad (\kappa=1,2) \quad ,$$

and analogously with y_{II} instead of y_I . Then one obtains holomorphic connections of the $me_{\nu\pm2n}$ with se_{2n} and ce_{2n} . This can be translated to the Fourier coefficients. But the arizing formulas are obviously not simple. Therefore we forgo the reproduction of the results.

It fits into the present context to make MS 2.25. Theorem 13 more precise. M.Bell (1957), D.M.Levy and J.B.Keller (1963) and H.Hochstadt (1964) give

$$a_m(h^2) - b_m(h^2) = \frac{2h^{2m}}{[2^{m-1}(m-1)!]^2} \quad (1 + \mathcal{O}(h^4)) .$$

3. Spheroidal Functions

3.1. Integrals with Products of Spheroidal Functions.

3.1.1. Integral relations of the first kind.

From

$$(1) \qquad \left[(1-z^2)y_\kappa'(z)\right]' + \left[\lambda_\kappa + \gamma^2(1-z^2) - \frac{\mu_\kappa^2}{1-z^2}\right] y_\kappa(z) = 0 \qquad (\kappa = 1,2)$$

there results by crosswise multiplication with $y_\kappa(z)$ $(\kappa = 1,2)$, subtraction and integration along a path from a to b that shall not meet $+1$ and -1

$$(2) \qquad \begin{aligned} &(z^2-1)(y_1'(z)y_2(z) - y_1(z)y_2'(z))'\Big|_a^b = \\ &= (\lambda_1-\lambda_2)\int_{\mathcal{L}} y_1(z)y_2(z)dz + \left(\mu_1^2 - \mu_2^2\right)\int_{\mathcal{L}} (z^2-1)^{-1}y_1(z)y_2(z)dz \,. \end{aligned}$$

If the left member is known, then one has with (2) a representation of certain integrals over products of spheroidal functions.

If, in particular, one assumes

$$\gamma^2 \neq 0, \ \nu_1 \not\equiv \tfrac{1}{2}(\text{mod } 1), \ \nu_2 \not\equiv \tfrac{1}{2}(\text{mod } 1)$$

and chooses

$$\lambda_1 = \lambda_{\nu_1}^{\mu_1}(\gamma^2) \ , \ \gamma_2 = \lambda_{\nu_2}^{\mu_2}(\gamma^2)$$

and, according to MS 3.64.,

$$y_1(z) = S_{\nu_1}^{\mu_1(1)}(z;\gamma) \ , \ y_2(z) = S_{\nu_2}^{\mu_2(1)}(z;\gamma)$$

with

$$\arg \gamma = -\alpha(\text{mod } 2\pi) \ , \quad -\pi < \alpha \leq \pi,$$

then one can derive with MS 3.65., (48), (49), (50)

$$(3) \qquad \begin{aligned} &\int_{\infty e^{i\alpha}}^{(+1+,-1+)} \left[\lambda_{\nu_1}^{\mu_1}(\gamma^2) - \lambda_{\nu_2}^{\mu_2}(\gamma^2) + \frac{\mu_1^2 - \mu_2^2}{z^2-1}\right] S_{\nu_1}^{\mu_1(1)}(z;\gamma)\, S_{\nu_2}^{\mu_2(1)}(z;\gamma)dz = \\ &= \left(e^{2\pi i(\nu_1+\nu_2)} - 1\right)\frac{1}{\gamma}\sin\frac{\nu_1-\nu_2}{2}\pi \ . \end{aligned}$$

Considering the relation in MS 3.65. (47)

$$\cos \nu\pi \cdot S_\nu^{\mu(2)}(z,\gamma) = -S_{-\nu-1}^{\mu(1)}(z;\gamma) - \sin \nu\pi \, . \, S_\nu^{\mu(1)}(z;\gamma) \ ,$$

one can evaluate integrals of the type (3) in which one or both spheroidal functions of the first kind are replaced by spheroidal functions of the second kind.

If, in addition,

γ^2 is not exceptional value to ν_κ $(\kappa = 1,2)$,

$-\pi < \alpha < \pi,\ \mu_1 = \mu_2 = \mu$,

then one can obtain in this manner, but also directly,

$$(4)\quad \left(\lambda^\mu_{\nu_1}(\gamma^2) - \lambda^\mu_{\nu_2}(\gamma^2)\right) \int\limits_{\infty e^{i\alpha}}^{(+1+)} Ps^\mu_{\nu_1}(z;\gamma^2) Ps^\mu_{\nu_2}(z;\gamma^2)dz = (e^{-2\pi i\mu}-1)C(\nu_1,\nu_2,\mu,\gamma^2)$$

with

$$\gamma \cos \nu_1\pi \cos \nu_2\pi \cdot C(\nu_1,\nu_2,\mu,\gamma^2) =$$

$$(5)\quad = \sin\frac{\nu_1-\nu_2}{2}\pi \left[\frac{1}{\Gamma(\nu_1-\mu+1)\Gamma(\nu_2-\mu+1)V^\mu_{\nu_1}(\gamma)V^\mu_{\nu_2}(\gamma)} - \frac{1}{\Gamma(-\nu_1-\mu)\ (-\nu_2-\mu)V^\mu_{-\nu_1-1}(\gamma)V^\mu_{-\nu_2-1}(\gamma)}\right] +$$

$$+ \sin\frac{\nu_1+\nu_2+1}{2}\pi \left[\frac{1}{\Gamma(-\nu_1-\mu)\Gamma(\nu_2-\mu+1)V^\mu_{-\nu_1-1}(\gamma)V^\mu_{\nu_2}(\gamma)} - \frac{1}{\Gamma(\nu_1-\mu+1)\ (-\nu_2-\mu)V^\mu_{\nu_1}(\gamma)V^\mu_{-\nu_2-1}(\gamma)}\right] .$$

For

$$\mathrm{Re}\ \mu < 1$$

one can rewrite (4) as

$$(6)\quad \left(\lambda^\mu_{\nu_1}(\gamma^2) - \lambda^\mu_{\nu_2}(\gamma^2)\right) \int\limits_1^{\infty e^{i\alpha}} Ps^\mu_{\nu_1}(z;\gamma^2) Ps^\mu_{\nu_2}(z;\gamma^2)dz = C(\nu_1,\nu_2,\mu,\gamma^2) .$$

Further special cases for specified values of the parameters can be easily discussed. See also D.R.Rhodes (1964), (1969).

We note finally a remarkable consequence. If one sets in (3) $\mu_1 = \mu_2 = \mu$, $\nu_2 = -\nu - 1$, divides by $\nu_1 - \nu$ and goes to the limit ν , then one obtains with MS. 3.66., (56), (58) and 3.543., (24).

$$(7)\quad \frac{\partial\lambda^\mu_\nu(\gamma^2)}{\partial\nu} = (2\nu + 1)\ A^\mu_\nu(\gamma^2)\ A^{-\mu}_\nu(\gamma^2) .$$

Of interest is the orthogonality property for $\gamma \in \mathbb{R}, n,\ell,m \in \mathbb{N}_0,\ n,\ell \geq m$.

$$(8)\quad \int\limits_{-\infty}^{\infty} [\mathrm{sign}(z^2-1)]^m\ ps^m_n(z)ps^{-m}_\ell(z)dz = \left\{1 - (-1)^m + \frac{\pi}{2\gamma A^m_n A^{-m}_n K^m_n K^{-m}_n}\right\} \frac{2}{2n+1}\ \delta_{n\ell} ,$$

which is also an immediate consequence of (2). For $|\ell - n| =$ odd, the principal value at $z = \pm\infty$ must be taken in the integral. This is a counterpart to the orthogonality over the interval $-1 \leq z \leq 1$ (see MS 3.23.)

$$(9) \qquad \int_{-1}^{1} (-1)^m \, ps_n^m(z) \, ps_\ell^{-m}(z) = \frac{2}{2n+1} \delta_{n\ell} \quad .$$

3.1.2. Integral relations of the second kind.

F.W.Schäfke (1957) has given a second kind of integral relations for spheroidal functions. A short review will now be given.

If $u(\xi,\eta,\varphi)$ is a holomorphic solution of the wave equation in prolate spheroidal coordinates ξ,η,φ in a domain of C^3 with $\xi \neq \pm 1$, $\eta \neq \pm 1$, $\xi \neq \pm\eta$, then the same property holds for

$$\frac{1}{\xi^2 - \eta^2}\left[\eta(\xi^2 - 1)\frac{\partial u}{\partial \xi} + \xi(1-\eta^2)\frac{\partial u}{\partial \eta}\right]$$

and

$$e^{\pm i\varphi}\left\{\frac{\sqrt{(\xi^2-1)(1-\eta^2)}}{\xi^2 - \eta^2}\left[\xi\frac{\partial u}{\partial \xi} - \eta\frac{\partial u}{\partial \xi}\right] \pm i \frac{1}{\sqrt{(\xi^2-1)(1-\eta^2)}} \frac{\partial u}{\partial \varphi}\right\} \quad .$$

For these expressions are the derivatives with respect to cartesian coordinates

$$c\frac{\partial u}{\partial x_3} \quad , \quad c\left(\frac{\partial u}{\partial x_1} \pm i \frac{\partial u}{\partial x_2}\right) \quad .$$

If this is applied to

$$u(\xi,\eta,\varphi) = S_\nu^{\mu(j)}(\xi;\gamma) \, \widetilde{Q}s_\nu^\mu(\eta;\gamma^2) \, e^{i\mu\varphi} , \qquad (j = 3,4)$$

then one obtains according to MS 1.133., Theorem 1 the following kernels for integral relations between spheroidal functions:

$$(1) \quad v_1(\xi,\eta) = \frac{1}{\xi^2-\eta^2}\left[\eta(\xi^2-1)S_\nu^{\mu(j)'}(\xi;\gamma)\widetilde{Q}s_\nu^\mu(\eta;\gamma^2) + \xi(1-\eta^2)S_\nu^{\mu(j)}(\xi;\gamma)\widetilde{Q}s_\nu^{\mu'}(\eta;\gamma^2)\right] ,$$

$$(2) \quad v_2(\xi,\eta) = \frac{(\xi^2-1)^{1/2}(\eta^2-1)^{1/2}}{\xi^2 - \eta^2}\left[\xi S_\nu^{\mu(j)'}(\xi;\gamma)\widetilde{Q}s_\nu^\mu(\eta;\gamma^2) - \eta S_\nu^{\mu(j)}(\xi;\gamma)\widetilde{Q}s_\nu^{\mu'}(\eta;\gamma^2)\right] -$$

$$- \frac{\mu}{(\xi^2-1)^{1/2}(\eta^2-1)^{1/2}} S_\nu^{\mu(j)}(\xi;\gamma) \, \widetilde{Q}s_\nu^\mu(\eta;\gamma^2) \; .$$

In v_1 a factor $e^{i\mu\varphi}$, in v_2 (upper sign) a factor $e^{i(\mu+1)\varphi}$ is separated off.

Now one considers a contour in the η - plane, which runs about +1 and -1 once in the positive sense, and interprets

$$(\xi^2 - 1)^{1/2} \sim \xi \qquad (\xi \to \infty)$$

$$(\eta^2 - 1)^{1/2} \sim \eta \qquad (\eta \to \infty)$$

outside $[-1,1]$ uniquely. Then one can consider the integrals

$$(3) \qquad \oint v_1(\xi,\eta) \, \widetilde{Q}s_{-\nu-p-1}^\mu(\eta;\gamma^2) d\eta$$

and

(4) $$\oint v_2(\xi,\eta)\ \widetilde{Qs}^{\mu+1}_{-\nu-p-1}(\eta;\gamma^2)\,d\eta$$

with integer p and obtains with MS 1.133., Theorem 1 spheroidal functions. They are easily identified by their asymptotic behavior and are

(5) $$\gamma(\pm i)^{p+1}\oint \eta\ \widetilde{Qs}^{\mu}_{\nu}(\eta;\gamma^2)\ \widetilde{Qs}^{\mu}_{-\nu-p-1}(\eta;\gamma^2)\,d\eta\cdot S^{\mu(j)}_{\nu+p}(\xi;\gamma)$$

and

(6) $$\gamma(\pm i)^{p+1}\oint (\eta^2-1)^{1/2}\ \widetilde{Qs}^{\mu}_{\nu}(\eta;\gamma^2)\ \widetilde{Qs}^{\mu+1}_{-\nu-p-1}(\eta;\gamma^2)\,d\eta\cdot S^{\mu+1(j)}_{\nu+p}(\xi;\gamma)\quad,$$

respectively, with the upper sign for $j = 3$, the lower sign for $j = 4$.

It is easy to recognize when these integrals vanish for certain values of the parameters, or when the contours can be cut in half, or when these integrals can be reduced to $(-1,1)$. We refrain from noting down the special cases with integer ν and μ.

If one writes down the integral relations (1), (3), (5), and (2), (4), (6), respectively, for $j = 3,4$ each, then one obtains linear systems of equations for the integrals

(7) $$\oint \frac{\eta}{\xi^2-\eta^2}\ \widetilde{Qs}^{\mu}_{\nu}(\eta;\gamma^2)\ \widetilde{Qs}^{\mu}_{-\nu-p-1}(\eta;\gamma^2)\,d\eta\quad,$$

(8) $$\oint \frac{1-\eta^2}{\xi^2-\eta^2}\ \widetilde{Qs}^{\mu\,\prime}_{\nu}(\eta;\gamma^2)\,\widetilde{Qs}^{\mu}_{-\nu-p-1}(\eta;\gamma^2)\,d\eta\quad,$$

(9) $$\oint \frac{(\eta^2-1)^{1/2}}{\xi^2-\eta^2}\ \widetilde{Qs}^{\mu}_{\nu}(\eta;\gamma^2)\ \widetilde{Qs}^{\mu+1}_{-\nu-p-1}(\eta;\gamma^2)\,d\eta\quad,$$

(10) $$\oint \frac{(\eta^2-1)^{1/2}\eta}{\xi^2-\eta^2}\ \widetilde{Qs}^{\mu\,\prime}_{\nu}(\eta;\gamma^2)\ \widetilde{Qs}^{\mu+1}_{-\nu-p-1}(\eta;\gamma^2)\,d\eta$$

which can be easily solved, and these integrals are in turn equal to

(11) $$\frac{\gamma^2}{2i}(-1)^{\frac{p+1}{2}}\oint \eta\ \widetilde{Qs}^{\mu}_{\nu}(\eta;\gamma^2)\ \widetilde{Qs}^{\mu}_{-\nu-p-1}(\eta;\gamma^2)\,d\eta\ \times$$
$$\times\left[S^{\mu(3)}_{\nu+p}(\xi;\gamma)S^{\mu(4)}_{\nu}(\xi;\gamma) - S^{\mu(4)}_{\nu+p}(\xi;\gamma)\ S^{\mu(3)}_{\nu}(\xi;\gamma)\right]\quad,$$

$$\frac{\gamma^2}{2i}(-1)^{\frac{p+1}{2}} \frac{\xi^2-1}{\xi} \oint \eta \, \widetilde{Qs}_\nu^\mu(\eta;\gamma^2)\, \widetilde{Qs}_{-\nu-p-1}^\mu(\eta;\gamma^2)\, d\eta \;\times \tag{12}$$

$$\times \left[S_{\nu+p}^{\mu(4)}(\xi;\gamma)\, S_\nu^{\mu(3)\prime}(\xi;\gamma) - S_{\nu+p}^{\mu(3)}(\xi;\gamma)\, S_\nu^{\mu(4)\prime}(\xi;\gamma) \right] ,$$

$$\frac{\gamma^2}{2i}(-1)^{\frac{p+1}{2}} \frac{(\xi^2-1)^{1/2}}{\xi} \oint (\eta^2-1)^{1/2}\, \widetilde{Qs}_\nu^\mu(\eta;\gamma^2)\, \widetilde{Qs}_{-\nu-p-1}^{\mu+1}(\eta;\gamma^2)\, d\eta \;\times \tag{13}$$

$$\times \left[S_{\nu+p}^{\mu+1(3)}(\xi;\gamma)\, S_\nu^{\mu(4)}(\xi;\gamma) - S_{\nu+p}^{\mu+1(4)}(\xi;\gamma)\, S_\nu^{\mu(3)}(\xi;\gamma) \right] ,$$

$$\frac{\gamma^2}{2i}(-1)^{\frac{p+1}{2}} (\xi^2-1)^{1/2} \int (\eta^2-1)^{1/2}\, \widetilde{Qs}_\nu^\mu(\eta;\gamma^2)\, \widetilde{Qs}_{-\nu-p-1}^{\mu+1}(\eta;\gamma^2)\, d\eta \;\times$$

$$\times \left[S_{\nu+p}^{\mu+1(3)}(\xi;\gamma)\, S_\nu^{\mu(4)\prime}(\xi;\gamma) - S_{\nu+p}^{\mu+1(4)}(\xi;\gamma)\, S_\nu^{\mu(3)\prime}(\xi;\gamma) \right] - \tag{14}$$

$$- \frac{\mu}{\xi^2-1} \int (\eta^2-1)^{1/2}\, \widetilde{Qs}_\nu^\mu(\eta;\gamma^2)\, \widetilde{Qs}_{-\nu-p-1}^{\mu+1}(\eta;\gamma^2)\, d\eta \quad .$$

Further interesting formulas arise from these integral relations by comparing the asymtotic behavior for $\xi \to \infty$.

3.2. On the eigenvalues for complex γ^2.

As in the case of Mathieu's differential equation it is also here desirable to obtain more detailed quantitative and qualitative knowledge of the Riemann surface or of the analytic functions in the large, respectively, beyond the results in MS 3.22., 3.24., 3.253., 3.53. .

In particular for the $\lambda_n^m(\gamma^2)$ with $m = 0,1,2,\ldots$ and

$$n = m,\ m+2,\ m+4,\ \ldots$$

and

$$n = m+1,\ m+3,\ m+5,\ \ldots \quad ,$$

respectively, F.W.Schäfke, R.Ebert, H.Groh and A.Schönhage , 1959 - 1962, have carried out extensive numerical-function-theoretical investigations by means of the electronic computer ER 56 of the University of Cologne. The approximate values (with significant figures, no rounding up) of a series of branch points are here given for the first time.

The following tables give the real and imaginary parts of γ^2 for these branch points. Their first columns contain the branches which are connected in radial continuation from $\gamma^2 = 0$ in a turn about the respective γ^2, indicated by the values of $\lambda_n^m(0) = n(n+1)$.

m = 0.	n - m = 0,2,4,...	
branches	Re γ^2	Im γ^2
0 → 6	-3,44	9,49
6 → 20	17,72	32,15
"	-29,35	24,33
20 → 42	-9.28	77,75
"	60,08	56,52
"	-75,49	40,28
42 → 72	35,26	132,96
"	-59,09	125,68
"	122,75	82,11
"	-141,58	56,95
72 → 110	-15,08	211,27
"	101,76	190,36
"	-130,18	175,63
"	205,48	108,63
"	-227,52	74,15
110 → 156	52,70	299,13

m = 0.	n - m = 1,3,5,...	
branches	Re γ^2	Im γ^2
2 → 12	4,36	20,58
12 → 30	-22,70	50,96
"	36,34	44,15
30 → 56	10,16	105,07
"	-71,22	83,37
"	88,90	69,18
56 → 90	-40,20	168,23
"	65,84	161,40
"	-140,28	117,30
"	161,62	95,27
90 → 132	15,95	254,90
"	-112,63	233,68
"	142,95	219,80
"	-229,55	152,41
"	254,32	122,18

m = 1	n - m = 0,2,4,...	
branches	Re γ^2	Im γ^2
2 → 12	-12,79	16,38
12 → 30	8,43	53,50
"	-48,90	32,00
30 → 56	-30,37	100,78
"	52,44	92,73
"	-105,04	48,41
56 → 90	14,23	170,73
"	-90,96	149,96
"	117,50	133,84
"	-181,07	65,39
90 → 132	-47,80	250,57
"	81,67	243,02
"	-172,46	200,89
"	203,08	176,46
"	-276,92	82,83
132 → 182	20,02	353,27

m = 1.	n - m = 1,3,5,...	
branches	Re γ^2	Im γ^2
6 → 20	-5,26	34,71
20 → 42	27,75	72,86
"	-43,30	66,51
42 → 72	-11,04	135,47
"	82,38	113,08
"	-102,17	99,86
72 → 110	45,19	206,59
"	-72,63	200,06
"	157,75	154,98
"	-181,37	134,48
110 → 156	-16,83	301,62

m = 2.	n - m = 0,2,4,...	
branches	Re γ^2	Im γ^2
6 → 20	-25,13	23,29
20 → 42	-4,59	74,23
"	-71,39	39,64
42 → 72	40,30	127,21
"	-54,71	123,45
"	-137,52	56,49
72 → 110	-10,45	207,62
"	107,06	182,47
"	-125,92	173,97
"	-223,49	73,79

m = 2.	n - m = 1,3,5,...	
branches	Re γ^2	Im γ^2
12 → 30	-18,23	48,63
30 → 56	15,03	100,42
"	-66,98	81,90
56 → 90	-35,68	165,27
"	71,02	154,58
"	-136,12	116,21
90 → 132	20,70	250,56
"	-108,27	231,48
"	148,37	210,84
"	-225,44	151,55

m = 3.	n - m = 0,2,4,...	
branches	Re γ^2	Im γ^2
12 → 30	-40,24	30,35
30 → 56	-20,91	94,99
"	-96,73	47,32
56 → 90	24,31	161,26
"	-82,16	146,05
"	-172,89	64,58
90 → 132	-38,55	244,16
"	92,25	230,03
"	-163,92	197,90
"	-268,80	82,18

m = 3.	n - m = 1,3,5,...	
branches	Re γ^2	Im γ^2
20 → 42	-34,22	62,69
42 → 72	-1,26	127,81
"	-93,62	97,30
72 → 110	55,53	195,34
"	-63,59	194,88
"	-173,02	132,53

m = 4.	n - m = 0,2,4,...	
branches	Re γ^2	Im γ^2
20 → 42	-57,99	37,59
42 → 72	-40,31	115,95
"	-124,84	55,07
72 → 110	4,82	195,30
"	-112,60	168,73
"	-211,07	72,69

m = 4.	n - m = 1,3,5,...	
branches	Re γ^2	Im γ^2
30 → 56	-53,07	76,99
56 → 90	-20,83	155,32
"	-123,13	112,79
90 → 132	36,33	235,95
"	-94,63	224,51
"	-212,81	148,89

m = 5.	n - m = 0,2,4,...	
branches	Re γ^2	Im γ^2
30 → 56	-78,32	45,00
56 → 90	-62,61	137,19
"	-155,65	62,90
90 → 132	-17,94	229,54
"	-145,96	191,54
"	-252,03	80,85

m = 5.	n - m = 1,3,5,...	
branches	Re γ^2	Im γ^2
42 → 72	-74,65	91,55
72 → 110	-43,50	183,08
"	-155,45	128,42

m = 6.	n - m = 0,2,4,...	
branches	Re γ^2	Im γ^2
42 → 72	-101,15	52,59
72 → 110	-87,71	158,74
"	-189,12	70,83

m = 6.	n - m = 1,3,5,...	
branches	Re γ^2	Im γ^2
56 → 90	-98,88	106,38
90 → 132	-69,11	211,16
"	-190,50	144,19

m = 7.	n - m = 0,2,4,...	
branches	Re γ^2	Im γ^2
56 → 90	-126,44	60,34
90 → 132	-115,53	180,61
"	-225,19	78,85

m = 7.	n - m = 1,3,5,...	
branches	Re γ^2	Im γ^2
72 → 110	-125,70	121,48

m = 8.	n - m = 0,2,4,...	
branches	Re γ^2	Im γ^2
72 → 110	-154,16	68,25

m = 8.	n - m = 1,3,5,...	
branches	Re γ^2	Im γ^2
90 → 132	-155,06	136,84

m = 9.	n - m = 0,2,4,...	
branches	Re γ^2	Im γ^2
90 → 132	-184,27	76,32

By more precise computations and error estimates the radii of convergence r_n^m of the first few λ_n^m were determined. In the following table approximate values of them are given with significant figures.

m = 0	n = 0/2	10,098478681412451544
	1/3	21,0352862295526809
	4	36,7128570517630
	5	55,784916999826
	6	78,298772832835
	7	105,5555790047
	8	137,55446800
	9	172,9641237
	10	211,8103161
	11	255,39920
m = 1	n = 1/3	20,7833503707054564
	2/4	35,102242838118878
	5	54,1618613082227
	6	77,961692744065
	7	105,252514163
	8	135,9195737859
	9	171,32614790
	10	211,474684
	11	255,08640
m = 2	n = 2/4	34,263300066033015
	3/5	51,93128512195512
	6	74,376058831558
	7	101,54190164829
	8	133,4444974366
	9	169,0779777
	10	207,87993494
	11	251,41774
m = 3	n = 3/5	50,401535606941214
	4/6	71,4231472402646
	7	97,261312781965
	8	127,813902030
	9	163,080611096
	10	203,0801085
	11	247,1871896
m = 4	n = 4/6	69,1099794169652
	5/7	93,505951330863
	8	122,75209003774
	9	156,7092465822
	10	195,363717814
	11	238,72842820
m = 5	n = 5/7	90,32600177384577
	6/8	118,125035495140
	9	150,7983135697
	10	188,17806396
	11	230,24036735
m = 6	n = 6/8	114,002241435105
	7/9	145,23718193889
	10	181,3598351240
	11	222,181084459
m = 7	n = 7/9	140,101337009119
	8/10	174,807188346
	11	214,403460702
m = 8	n = 8/10	168,59288920197
	9/11	206,805707671
m = 9	n = 9/11	199,45157056353

This table shows - as in 2.4. - that also here the radii of convergence seem to grow according to a quadratic law, contrary to the linear estimate from below of the perturbation theory. This leads to the conjecture

$$r_-^m = a_- n^2 + b_- n + c_- + \mathcal{O}(n^{-1}) \quad .$$

In particular, one has approximately

$$a_o = 2{,}042,\ b_o = 0{,}71, \qquad 0 < c_o < 1{,}5\ ;$$

Also for $m \neq 0$ it seems that

$$a_m = 2{,}042\ ;$$

see also 2.4. and the conjecture of the coefficient 2,042.

Finally we note according to F.W.Schäfke (1975) the

Theorem : The function elements about $\gamma^2 = 0$

$$\lambda_\nu^\mu(\gamma^2) \qquad (\nu=\mu,\ \mu+2,\ \mu+4,\ldots)$$

and

$$\lambda_\nu^\mu(\gamma^2) \qquad (\nu=\mu+1,\ \mu+3,\ \mu+5,\ldots)\ ,$$

respectively, belong for real, non-negative, not half of an odd integer and not too large values of μ (certainly for $0 \leq \mu \leq 6$) each to one analytic function in the large.

The proof is almost word for word the same as for the corresponding theorem in 2.4., but it uses the coefficients of γ^6.

3.3. The spheroidal functions for $\mu^2 = 1$, $\lambda = 0$.

The transformation of MS 3.14., (7), (8) shows, that the spheroidal differential equation

$$[(1-z^2)y'(z)]' + [\lambda + \gamma^2(1-z^2) - \mu^2(1-z^2)^{-1}]y(z) = 0$$

for $\mu^2 = 1$ and $\lambda = 0$ has the solutions

(1) $$y(z) = (z^2-1)^{-1/2}\, e^{\pm i\gamma z}\ ;$$

see also MS 3.11., Satz 5, MS 3.534. and MS 3.65. .

For

(2) $$\gamma^2 = \gamma_n^2 := n^2\left(\frac{\pi}{2}\right)^2 \qquad (n=1,2,3,\ldots)$$

the functions

(3) $$\begin{aligned} y(z) &= (1-z^2)^{-\frac{1}{2}} \sin\gamma_n(1-z) \\ &= (-1)^{\frac{n-1}{2}} (1-z^2)^{-\frac{1}{2}} \cos\gamma_n z \qquad (n \text{ odd}) \\ &= (-1)^{\frac{n}{2}-1} (1-z^2)^{-\frac{1}{2}} \sin\gamma_n z \qquad (n \text{ even}) \end{aligned}$$

belong for $z = +1$ and $z = -1$ to the index $+\frac{1}{2}$. Therefore the values (2) give the points with

(4) $$\lambda_n^1\left(\gamma_n^2\right) = 0 = \lambda_0^1\left(\gamma_n^2\right) \qquad (n=1,2,3,\ldots)\ ;$$

for, the number of zeros in the open interval $-1 < z < 1$ is $n - 1$; see MS 3.23., Satz 5.

These points $\lambda = 0$, $\gamma^2 = \gamma_n^2$ for $\mu^2 = 1$ have some interest: they give exceptional values for the eigenvalue problem of MS 3.51., by (4), but are on the other hand normal values for the eigenvalue problem MS 3.21.. Therefor we note some observations about the corresponding spheroidal functions and coefficients of expansions.

First, we choose $\mu^2 = 1$, $\lambda = 0$, $\gamma^2 \neq \gamma_n^2$. Then, according to MS 3.541., 3.542.

$$\alpha^{\pm 1}_{o,2r}(\gamma^2) = \alpha^{\pm 1}_{-1,2r}(\gamma^2) = 0 \qquad (r > 0)$$

and by

$$\alpha^{\pm 1}_{o,o}(\gamma^2) = \alpha^{\pm 1}_{-1,o}(\gamma^2) = 1$$

the normalization MS 3.542., (21) and Satz 1 with the definition of

$$\widetilde{Qs}^{\pm 1}_o(z;\gamma^2) \ , \ \widetilde{Qs}^{\pm 1}_{-1}(z;\gamma^2)$$

is possible. By MS 3.61., (4), (5) there holds

(5) $$ps^1_o(z;\gamma^2) = 0$$

MS 3.62. gives

(6) $$a^1_{o,o}(\gamma^2) = 1$$

and

(7) $$a^1_{o,2r}(\gamma^2) = 0 \qquad (r \neq 0),$$

and by MS 3.62., (7) again (5). MS 3.64., (41*), (42) yield

(8) $$A^1_o(\gamma^2) = 1$$

and, according to (1),

(9) $$s^{1(j)}_o(z;\gamma) = (z^2-1)^{-\frac{1}{2}} z \, \psi^{(j)}_o(\gamma z).$$

Let now $\gamma^2 = \gamma_n^2$. Then in MS 3.541., 3.542. for n odd (3) gives

$$\alpha^1_{o,2r} = 0 \qquad (r \geqq 0) \ ,$$

$$\alpha^1_{-1,2r} = 0 \qquad (r > 0) \ ,$$

for n even

$$\alpha^1_{-1,2r} = 0 \qquad (r \geqq 0) \ ,$$

$$\alpha^1_{o,2r} = 0 \qquad (r > 0) \ .$$

Therefore normalization with MS 3.542., (21), (24*) is impossible; the functions

$$\widetilde{Qs}^{\pm 1}_o(z;\gamma^2) \ , \ \widetilde{Qs}^{\pm 1}_{-1}(z;\gamma^2)$$

and the functions MS 3.61. are not defined. But according to MS 3.2.

(10) $$ps_n^1(z;\gamma_n^2) = \sum_{n+2r \geq 1}^{\infty} b_{n,2r}\, P_{n+2r}^1(z) \qquad (n=1,2,3,\ldots)$$

are defined with

(11) $$b_{n,2r} = \lim_{\gamma^2 \to \gamma_n^2} a_{n,2r}^1(\gamma^2).$$

Here, for $\gamma^2 \neq \gamma_n^2$, $\left(a_{n,2r}^1(\gamma^2)\right)_{r \in \mathbb{Z}}$

is normed solution of MS 3.62., (10) with $\mu = 1$, $\nu = n$. One has

$$a_{n,2r}^1(\gamma^2) = 0 \qquad (n+2r < -1);$$

but with

$$n + 2r_o = \begin{cases} 0 & (n \text{ even}) \\ -1 & (n \text{ odd}) \end{cases}$$

there holds

$$\gamma^2 \frac{(n+2r_o+3)(n+2r_o+2)}{(2n+4r_o+3)(2n+4r_o+5)} a_{n,2r_o+2}^1(\gamma^2) + \lambda_n^1(\gamma^2)\, a_{n,2r_o}^1(\gamma^2) = 0 .$$

This shows, that for $\gamma^2 \to \gamma_n^2$ the coeffizient $a_{n,2r_o}^1(\gamma^2)$ has a simple pole; this also holds for $A_n^1(\gamma^2)$. Further it is obvious, in this connection, that the coefficients (11) are not part of a solution of MS 3.6 2., (10). For $\gamma^2 = \gamma_n^2$ the solution of MS 3.62., (10), which is required in MS 3.64. for $S_n^{1(j)}(z;\gamma)$, is, according to (6), (7), (8),

$$a_{n,2r_o}^1(\gamma_n^2) = 1 , \quad a_{n,2r_o}^1(\gamma_n^2) = 0 \qquad (r \neq r_o)$$

and with it

$$A_n^1(\gamma_n^2) = (-1)^n .$$

This, in MS 3.64., leads to

(12) $$S_n^{1(j)}(z;\gamma_n) = (z^2-1)^{-\frac{1}{2}}\, z(-1)^n\, \psi_{n+2r_o}^{(j)}(\gamma_n z) ,$$

according to (9) and (1). The behaviour of the $a_{n,2r}^1(\gamma^2)$, $A_n^1(\gamma^2)$, obtained for $\gamma^2 \to \gamma_n^2$, makes obvious that

(13) $$S_n^{1(j)}(z;\gamma) \to S_n^{1(j)}(z;\gamma_n)$$

in MS 3.64., (42). -

Similar situations hold for $m = 2,3,4,\ldots$ and the intersections of $A_m(\lambda,\gamma^2) = 0$ with $\lambda = \lambda_n^m(\gamma^2)$ $(n \geq m)$, especially for the real points which occur for $m = 3,5,7,\ldots$; see also figure 16, MS, p.237, in case $m = 3$. For $m = 1$ see also L.Robin (1966).

As an illustration of the corresponding facts about the expansions for exceptional points λ^2 in 1.1., we note, that for $\mu^2 = 1$, $\lambda = 0$, $\lambda^2 \neq 0$ one has principal solutions of

$$[(1-z^2)y'(z)]' + [\gamma^2(1-z^2) - (1-z^2)^{-1}]y(z) = (z^2-1)^{-1/2} \cos \gamma z$$

and

$$[(1-z^2)\eta'(z)]' + [\gamma^2(1-z^2) - (1-z^2)^{-1}]\eta(z) = (z^2-1)^{-1/2} \sin \gamma z$$

with

$$y(ze^{\pi i}) = -y(z)$$

and

$$\eta(ze^{\pi i}) = \eta(z) ,$$

if and only if $\gamma^2 = \gamma_n^2$ $(n=1,2,3,\ldots)$. This is seen directly by the transformations

$$y(z) = -(z^2-1)^{-1/2}u(z) ,$$

$$\eta(z) = -(z^2-1)^{-1/2}v(z) ,$$

which lead to

(*) $$u''(z) + \gamma^2 u(z) = \frac{\cos \gamma z}{z^2 - 1} ,$$

(**) $$v''(z) + \gamma^2 v(z) = \frac{\sin \gamma z}{z^2 - 1}$$

with

$$u(ze^{\pi i}) = u(z) ,$$

$$v(ze^{\pi i}) = -v(z).$$

If $\gamma = \gamma_n$, n odd, in (*) the right member gives an even integer function and a suitable u obviously exists, while for the second problem one considers

$$w(z) = \cos \gamma_n z \cdot \log \frac{z - 1}{z + 1} ,$$

$$w''(z) + \gamma_n^2\, w(z) = \frac{w_1(z)}{z^2 - 1}$$

with w_1 odd and integer and $w_1(1) \neq 0$, which gives by suitable combination with (**) the existence of a solution of the second problem. Analogously for n even. That for $\gamma^2 \neq \gamma_n^2$ $(n=1,2,3,\ldots)$ the problems have no solution, is seen by a similar consideration as given for (**) above.

3.4. Applications and numerical tables.

During the last 25 years many new applications of spheroidal function have been made. Many of them are related to the wave equation and to generalizations of it. We mention diffraction by elliptical cylinder and by spheroids, vibrations of and radiation from such bodies. There are also many relevant papers in hydrodynamics and quantum-mechanics.

The prolate spheroidal functions $ps_n^o(z;\gamma^2)$ of order zero have found widespread interest in quite a different context, namely as solutions of the integral equation - see MS 3.83. -

$$\int_{-1}^{1} e^{i\gamma\xi\eta}\, ps_n(\eta)d\eta = 2i^n \alpha_n(\gamma) ps_n(\xi)\;(\gamma,\xi \in \mathbb{R})$$

and its iterate

$$\int_{-1}^{1} \frac{\sin \gamma(\xi-\eta)}{\gamma(\xi-\eta)}\, ps_n(\eta)d\eta = 2\alpha_n(\gamma)^2\, ps_n(\xi) \; .$$

According to the first integral equation, the prolate spheroidal functions are eigenfunctions of the finite Fourier transform, and no other eigenfunctions exist. The eigenvalues are given by

$$\alpha_n = A_n^0(\gamma^2) K_n^0(\gamma) \; .$$

Both integral relations can be inverted to yield

$$\frac{\gamma\alpha_n}{\pi}\, i^n \int_{-\infty}^{\infty} e^{-i\gamma\xi\eta}\, ps_n(\xi)d\xi = \begin{cases} ps_n(\eta) & (-1 \leqq \eta \leqq 1) \; , \\ 0 & (\eta \in \mathbb{R}, |\eta| > 1) \; ; \end{cases}$$

$$\int_{-\infty}^{\infty} \frac{\sin \gamma(\xi-\eta)}{\xi-\eta}\, ps_n(\xi)d\xi = \pi ps_n(\eta) \; .$$

Incidentally Parseval's equation for the Fouriertransform gives

$$\int_{-1}^{1} ps_n(\xi) ps_m(\xi)d\xi = \frac{2}{\pi}\gamma\alpha_n^2 \int_{-\infty}^{\infty} ps_n(\xi) ps_m(\xi)d\xi$$

- see also 3.1.1. (8) - .

By simple transformations the integration between -1 and +1 can be changed into any finite interval or band. Therefore, the correspondingly transformed prolate spheroidal functions are a set of bandlimited (or time limited as the case may be in an application) functions which are orthogonal and complete over the respective interval. They are also orthogonal, but not complete over the infinite interval.

The above mentioned properties of the prolate spheroidal functions are at the base of a great variety of applications which have been studied since 1954. We mention here stochastic processes (D.Slepian (1954)), laser modes (G.D.Boyd and J.P.Gordon (1961)) , modified versions of the uncertainty principle (H.J.Landau and H.O. Pollak (1961)), antenna theory (D.R.Rhodes (1963)), problems in communication theory (see the mentioned papers by Slepian and by Landau and Pollak, but also M.Petrich (1963)). But there are also such seemingly unconnected fields as the distribution of eigenvalues of a stochastic matrix which lead to the prolate spheroidal functions (M.Goudin (1961) , J.des Cloizeaux and M.Z.Mehta (1973)). For many applications of prolate spheroidal functions to laser modes, extrapolation of image date, extrapolation beyond the optical bandwith, degrees of freedom in the image, evaluation of wave aberrations, we refer to the article by B.R.Frieden (1971).

It should be mentioned that in some of these applications a natural generalization of the spheroidal functions plays an important part. They are solutions of the differential equation

$$\frac{d}{dz}(z^2-1)\frac{dy}{dz}+\left[-\Lambda-\frac{\mu^2}{z^2-1}+\frac{\lambda(\lambda+1)}{z^2}-\gamma^2+\gamma^2z^2\right]y=0\,. \qquad (*)$$

This differential equation has been considered, perhaps for the first time, by R.S.B.Palero (1956). Many properties of its solutions have been derived by A.Leitner and J.Meixner (1959), (1960). Further properties have beeen given by D.Slepian (1964) and J.C.Heurtley (1964), (1965) although only for $\mu = 0$.

A remarkable feature is the fact that the differential equation (*) has not only the symmetry $z \to -z$, but also another symmetry: To any solution $y = f(\Lambda,\mu,\lambda,\gamma,z)$ there exist solutions

$$y = z^{1/2}(z^2-1)^{-1/4}f\left(\Lambda+\gamma^2,\pm(\lambda+\tfrac{1}{2}),\pm\,\mu-\tfrac{1}{2},\pm\, i\gamma,(1-z^2)^{1/2}\right)$$

with independent $\pm$ signs.

The methods applied in the theory of spheroidal functions can used in this case, too. Also the theory of two parameter eigenvalue problem can be applied (set μ or λ constant).

Another interesting application is in nuclear theory (W.E.Frahn and R.H.Lemmer (1957)).

There are other generalizations of the spheroidal function. Various papers by R.K.Gupta (1975), (1977) and also C.A.Coulson and P.O.Robinson (1958) should be mentioned in this context. A comprehensive presentation is given by I.V.Komarov, L.I.Ponomarev and S.Ju.Slavyanov (1976).

Aids in the numerical computation of spheroidal functions are the various known expansions. In particular, there exist now many new results on γ - asymptotics. We mention, in particular, the papers by R.B.Dingle and H.J.W.Müller (1964), by H.J.W.Müller (1963), (1964), (1965), by D.Slepian (1965), by J.des Cloizeaux and M.L.Mehta (1972), by J.W.Miles (1975), and by S.Jorna and C.Springer (1971).

Of great help are the extensive tables of spheroidal functions by A.L.van Buren et.a.. (1975) and by S.Hanish et.al. (1970).

The first set of tables presents the prolate and the oblate angular spheroidal functions $ps_n^0(\cos\theta;\gamma^2)$ with $\theta = 0(1^o)\,90^o$, $n = 0(1)\,49$ and γ or $i\gamma$, respectively, = 0,1(0,1)1,0 (1)10(2)30(5)40 . In all cases 8 significant figures are given. A table of the associated eigenvalues, also with 8 significant figures, is appended. The user of these tables should be aware that the values of the oblate functions must be multiplied by $(-1)^{n(n-1)/2}$ in order that they agree in the limit $\gamma \to 0$ with the respective Legendre polynomials and prolate functions.

The second set of tables contains eigenvalues, the spheroidal functions of the first and second kind and their first derivative. The definition and normalization correspond to MS 3.65.(42) with $j = 1$ and $j = 2$. Eigenvalues, the spheroidal functions of the first kind and their first derivative are given to 18 significant figures. For the other functions 18 figures are given but in many cases less than 18 figures, in some cases only two, are significant. This is indicated by the accuracy index ACC in the tables, which just gives the number of significant figures.

The range of the variable and of the parameter is $m = 0,1,2$; $n = m(1)\ m + 49$; γ or $i\gamma$, respectively, $= 0{,}1(0{,}1)1(1)10(2)30(5)40$;
prolate functions: $\xi = 1 + 10^{-n}(n=8,7,6,5,4,3,2)$, $1{,}02(0{,}02)1{,}2(0{,}2)2(2)10$;
oblate functions: $-i\xi = 0(0{,}02)0{,}1(0{,}1)1{,}0(0{,}2)2{,}0$.

For available documentation on the used computer programs and for further results reference is made to A.L.van Buren, R.V.Baier, S.Hanish and B.J.King (1971).

Appendix

Correction of errors in MS.

p. 80 last line: read $(f,y_n^*(\overline{\mu}))y_n(\mu)$ instead of $(f,y_n(\overline{\mu}))y_n^*(\mu)$.

p. 96 line 12: read $\hat{C}_m$ instead of C_m.

after line (12) introduce:

$$\text{mit } \hat{C}_m = \sum_{s=-\infty}^{\infty} d_s(-1)^s(\nu+s,m).$$

p. 97 line before <u>1.93.</u>: attach a factor $i^{s\mp s}$ at the right end of the formula.

p. 115 equ. (22): read $c_{2\kappa-2r}^{\nu+2r}(h^2)\ c_{2\kappa-2s}^{\nu+2s}(h^2)$

instead of $c_{2\kappa}^{\nu+2r}(h^2)\ c_{2\kappa}^{\nu+2s}(h^2)$.

p. 121 line 4: replace the right member by

$$\frac{2h^{2m}}{[2^{m-1}(m-1)!]^2} + \mathcal{O}(h^{2m+2})$$

line 16: replace [3] by [4].

line 17: omit 52.

p. 126 line 2: read <u>1.28.</u> instead of <u>1.2.8.</u>.

p. 132 Abb. 6: exchange a_1 and b_1, also a_2 and b_2.

p. 138 line 6 from bottom: read zugleich instead of sogleich.

p. 139 line 2: y_m'' instead of y''.

p. 145 line 5: omit sin z.

p. 165 line 6 from bottom: read <u>2.32.</u> instead of <u>2.23.</u>.

p. 184 line 11: read u instead of w.

p. 190 equ. (14): read ce_m' instead of $\dot{ce}_m$.

p. 226 line 10: read $\cos\pi\sqrt{\lambda+\frac{1}{4}}$ instead of $\cos\sqrt{\lambda+\frac{1}{4}}$.

p. 233 line 6: read Satz 5 instead of Satz 15.

p. 240 line 12: read <u>1.65.</u> instead of <u>1.66.</u>.

p. 243 line 8 from bottom: read $\frac{1}{\gamma^5}[$ instead of $[\frac{1}{\gamma^5}$.

line 6 from bottom: read m^6 instead of m^2.

p. 246 line 5: read $\frac{2}{2n+1}$ instead of $\frac{1}{zn+1}$.

line 9: the same as in line 5.

p. 247 line 2: again the same as in p. 246 line 5 .

p. 260 equ. (2): read $z^{\nu-\mu}$ instead of $z^{\nu+\mu}$.

equ. (5): read $\binom{\mu}{t}$ instead of $\binom{u}{t}$.

p. 283 line 5: read $\tilde{\mathcal{Q}}^{\mu}_{\nu+2r}(z)$ instead of $\tilde{\mathcal{Q}}^{\mu}_{\nu+2r}(z;\gamma^2)$.

p. 286 after equ. (15): read Satz 3 und 3.542. instead of Satz 3.

p. 292 line 2 from bottom: read $\mu(3)$ instead of $(\mu 3)$.

p. 294 line 16: read $\nu = n = 0,1,\ldots m-1$, instead of $\nu = n = 0,1,2,\ldots$.

p. 296 bottom line: read c^*_s instead of c^* .

p. 298 line 14: read $-\frac{\mu}{2}$ instead of $-\mu$.

p. 299 bottom line: read $(\pm i0;\gamma)$ instead of $(\pm i0;\gamma^2)$; p.300 line 2: same.

p. 307 line 16: read Sphäroidkoordinaten instead of Sphäroidwellen.

line 19: introduce] before $b^{\mu}_{\nu,r}(\gamma;\alpha)$.

p. 309 line 12 from bottom: read $|\alpha \pm 1|$ instead of $|\alpha| \pm 1|$.

p. 318 line 6: read $\frac{1}{z+1} - \frac{3}{2}$ instead of $\frac{1}{z+1} - 2$.

p. 322 line 11 from bottom: introduce $(-1)^r$ after the first sum sign.

line 8 from bottom: introduce $(-1)^r$ after the first sum sign.

p. 327 line 5 from bottom: read 2.25. instead of 2.55. .

p. 331 equ. (19): read my'' instead of my .

p. 342 line 13: read $|n_1|^{+1/2}$ instead of $|n_1|^{-1/2}$.

p. 349 the text of 1. after (18) should be replaced by:

Mit den in 4.31., (5) und (6) eingeführten Abkürzungen und den Überlegungen zu 4.32., (8) erhält man so Lösungstypen der Form

$$(19a) \quad u = \sum_{m=0}^{\infty} \left(A_m Ce_m(\xi;h^2) ce_m(\eta;h^2) + B_m Ce_m(\xi;-h^2) ce_m(\eta;-h^2) \right) ,$$

$$(19b) \quad u = \sum_{m=0}^{\infty} \left(A_m Se_{m+1}(\xi;h^2) se_{m+1}(\eta;h^2) + B_m Se_{m+1}(\xi;-h^2) se_{m+1}(\eta;-h^2) \right) .$$

In Korrektur eines Irrtums bei McLachlan [5,7] erkennt man jedoch, daß im Gegensatz zu 4.31. hier keine endliche Linearkombination der Form (19a) oder (19b) eine nicht-triviale Lösung für die am Rand $(\xi = \xi_o)$ eingespannte Platte,

(20) $$u(\xi_0,\eta) = \frac{\partial u}{\partial \xi}(\xi_o,\eta) = 0 \quad ,$$

liefern kann. Je endlich viele der Funktionen $ce_m(\eta;h^2)$, $ce_m(\eta;-h^2)$, analog für $se_{m+1}(\eta;h^2)$, $se_{m+1}(\eta;-h^2)$, sind nämlich linear unabhängig, wie man leicht mit Hilfe der verschiedenen Differentialgleichungen nachweist. Hätte man also z.B. (20) für eine endliche Summe der Form (19a), so müßte

$$A_m Ce_m(\xi_0;h^2) = A_m Ce_m'(\xi_0,h^2) = 0 \quad ,$$

also

$$A_m = 0$$

gelten, analog $B_m = 0$. Entsprechendes gilt für (20) und (19b) und z.B. auch für die frei schwingende Platte:

$$\frac{\partial^2 u}{\partial \xi^2}(\xi_0,\eta) = \frac{\partial^3 u}{\partial \xi^3}(\xi_0,\eta) = 0 \quad .$$

Für diese Probleme bietet also die Theorie der Mathieuschen Funktionen im Gegensatz zu 4.31. keinen einfachen Zugang.

p. 357 line 12 from bottom: read u instead of $\underline{E}$ (twice).

p. 391 line 5: read $|m+n|$ instead of $|m+u|$.

---

p. 152 : for $\nu = \pm 4$ formula (25**) must be replaced by

$$\left.\begin{matrix} a_4 \\ b_4 \end{matrix}\right\} = 16 + \frac{128}{15} d^2 + \left\{\begin{matrix} \dfrac{256 \cdot 553}{3375} \\ -\dfrac{256 \cdot 197}{3375} \end{matrix}\right\} d^4 + \ldots \quad .$$

p. 175 line 15: $M_{\nu+s}^{(4)}$ instead of $M_{\nu+s}^{(j)}$.

p. 205 line 8: Se_{2n+2} instead of Se_{2n+1} .

BIBLIOGRAPHY

ALY,H.H., H.J.W.MÜLLER-KIRSTEN and N.VAHEDI-FARIDI: Scattering by singular potentials with a perturbation-theoretical introduction to Mathieu functions. J.Math. Phys. 16, 961-970 (1975).

AOI,T.: The steady flow of a viscous fluid past a fixed spheroidal obstacle at small Reynolds numbers. J.Phys.Soc.Japan 10, 119-129 (1955).

AOI,T.: On Spheroidal Functions. J.Phys.Soc.Japan, 10, 130-141 (1955).

ARSCOTT,F.M.: Periodic differential equations. An introduction to Mathieu, Lamé, and allied functions. Intern.Series of Monographs in Pure and Applied Mathematics, Vol.66. The Macmillan Co., New York, 1964.

ARSCOTT,F.M., and B.D.SLEEMAN: Multiplicative solutions of linear differential equations. J.London Math.Soc. 43, 263-270 (1968).

AUBERT,M., and N.R.G.BESSIS: Prolate spheroidal orbitals for homonuclear and heteronuclear diatomic molecules. I. Basic procedure. Phys.Rev.A 10, 51-60 (1974). II. Shielding effects for the two electron problem. Phys.Rev.A 10, 61-70 (1974).

BAIER,R.V.: Acoustic radiation impedance of caps and rings on oblate spheroidal baffles. J.Acoust.Soc. Anm. 51, 1705-1716 (1972).

BARAKAT,R.: Finite range integral equations with band-limited displacement kernels. Internat.J.Control (1) 15, 587-594 (1972).

BARAKAT,R., A.HOUSTON and E.LEVIN: Power series expansions of Mathieu functions with tables of numerical results. J.Math. and Phys. 42, 200-247 (1963).

BARCILON,V.: An isoperimetric problem for entire functions. Applicable Anal. 5, 109-115 (1975).

BELKINA,M.G.: Asymptotic representations of spheroidal functions with an azimuth index m = 1. Dokl.Akad.Nauks SSSR(N.S.) 114, 1185-1188 (1957).

BELL,M.: A note on Mathieu functions. Proc.Glasgow Math.Assoc. 3, 132-134 (1957).

BESSON,H.: Sur les fonctions spéciales qui apparaissent dans les représentations du groupe des mouvements du plan. C.R.Acad.Sci., Paris 273, 102-104 (1971).

BLANCH,G.: The asymptotic expansions for the odd periodic Mathieu functions. Trans. Amer.Math.Soc. 97, 357-366 (1960).

BLANCH,G.: Mathieu functions. p.721-750 in: Handbook of Mathematical Functions. Ed. by M.Abramowitz and I.A.Stegun. Dover Pub.,Inc., New-York, 1965.

BLANCH,G.: Numerical aspects of Mathieu eigenvalues. Rand.Circ.Matem.Palermo (II) 15, 51-97 (1966).

BLANCH,G.: Double points of Mathieu's equation. Math.Computation 23, 105-107 (1969).

BLANCH,G., and D.S.CLEMM: Tables relating to the radial Mathieu functions. Vol 1: Functions of the first kind (1962). Vol.2: Functions of the second kind (1965). U.S.Govt.Printing Office, Washington, D.C., 20402.

BLANCH,G., and D.S.CLEMM: Mathieu's equation for complex parameters. Tables of characteristic values. U.S.Govt. Printing Office, Washington,D.C. 20402. (1969).

BLANCH,G., and I.RHODES: Table of characteristic values of Mathieu's equation for large values of the parameter. J.Washington Acad.Sciences 45, No.6, June 1955 .

BOUWKAMP,C.J.: Theoretical and numerical treatment of diffraction through a circular aperture.. IEEE Trans.AP - 18, 152-176 (1970).

BOYD,G.D., and H.KOGELNIK: Generalized confocal resonator theory. Bell System.Techn. J. 41, 1347-1369 (1962).

BRESSER,H.: Reihenentwicklungen analytischer Funktionen nach Whittakerschen Funktionen $M_{\kappa,\nu+n}(z)$ und $M_{\kappa(\nu+n),\nu+n}((\nu+n)z)$ für $\nu = 0$ und $\nu = \frac{1}{2}$. Dissertation Köln 1962.

BUCK,G.J., and J.J.GUSTINSIC: Resolution limitations of a finite aperture. IEEE Trans.AP - 15, 376-381 (1967).

BÜHRING,W.: Schrödinger equation with inverse fourth-power potential, a differential equation with two irregular singular points. J.Mathem.Phys. 15, 1451-1459 (1974).

BURKE,J.E.: Approximations for some radial Mathieu functions. J.Math. and Phys. 43, 234-240 (1964).

BURKE,J.E.: Low frequency scattering by soft spheroids. J.Acoust.Soc.Amer. 39, 826-831 (1966).
BURKE,J.E.: Scattering by penetrable spheroids. J.Acoust.Soc.Am. 43, 871-875 (1968).
CALLAHAN,W.R.: Flexural vibrations of elliptical plates when transverse shear and rotary inertia are considered. J.Acoust.Soc.Am. 36, 823-829 (1964).
CAMPBELL,R.: Théorie gênérale de l'équation de Mathieu. Paris, 1955.
CHAKO,N.: On integral relations involving products of spheroidal functions. J.Math. and Phys. 36, 62-73 (1957).
CHANG,C.T.M.: Natural resonant frequency of a prolate acoustical resonator. J.Acoust. Soc.Am. 49, 611-614 (1971).
CHERTOCK,G.: Sound radiation from prolate spheroids. J.Acoust.Soc.Am. 33, 871-876 (1961).
COTTRELL,A.H., and M.A.JASWON: Distributions of solute atoms around a slow dislocation. P roc.Roy.Soc.A. 199, 104-114 (1949).
COULSON,C.A.,and P.D.ROBINSON: Wave functions for the hydrogen atom in spheroidal coordinates. I: The derivation and properties of the functions. Proc.Phys.Soc. 71, 815-827 (1958).
DAMBURG,R., and R.PROPIN: Über asymptotischc Entwicklungen oblater Sphäroidfunktionen und ihrer Eigenwerte. J.Reine Angew.Math. 233, 28-36 (1968).
DAYMOND,S.D.: The principal frequencies of vibrating systems with elliptic boundaries. Quart.J.Mech. and Appl.Math. 8, 361-372 (1955).
DES CLOIZEAUX,J., and M.L.MEHTA: Some asymptotic expressions for prolate spheroidal functions and for the eigenvalues of differential and integral equations of which they are solutions. J.Mathematical Phys. 13, 1745-1754 (1972).
DES CLOIZEAUX,J., and M.L.MEHTA: Asymptotic behavior of spacing distributions for the eigenvalues of random matrices. J.Mathematical Phys. 14, 1648-1650 (1973).
DINGLE,R.B., and H.J.W.MÜLLER: Asymptotic expansions of Mathieu functions and their characteristic numbers. J.reine angew. Math. 211, 11-32 (1962).
DINGLE,R.B., and H.J.W.MÜLLER: The form of the coefficients of the late terms in asymptotic expansions of the characteristic numbers of Mathieu and spheroidal-wave functions. J.reine angew.Math. 216, 123-133 (1964).
DÖRR,J.: Mathieusche Funktionen als Eigenfunktionen gewisser Integralgleichungen. Z.angew.Math.Mech. 38, 171-175 (1958).
EINSPRUCH,N.G., and C.A.BARLOW,JR.: Scattering of a compressional wave by a prolate spheroid. Quart.Appl.Math. 19, 253-258 (1961).
ERAŠEVSKAJA,S.P., and A.A.PAL'ČEV: The computation of spheroidal functions and their first derivatives on a computer. II.Vesci Akad.Navuk BSSR Ser.Fiz.-Mat.Navuk 1969, no.4, 37-46.
ERDÉLYI,A.,Ed.: Higher transcendental functions, Vol. III. McGraw-Hill Book Co.,Inc.; New York, Toronto, London 1955.
EROFEENKO,V.T.: Formulas for the mutual expansion of cylindrical and spheroidal wave functions. Differencial'nye Uravnenija 14, 915-918, 958 (1978).
EWIG,C.S., and D.O.HARRIS: Matrix solution of periodic Mathieu equations. J.Computational Phys. 11, 606-611 (1973).
FLAMMER,C.: Spheroidal wave functions. Stanford Univ.Press, Stanford,Calif., 1957.
FLETCHER,A., J.C.P.MILLER and L.ROSENHEAD: An index of mathematical tables. Vol. I. Introduction. Part I: Index according to functions Vol. II. Part II: Bibliography. Part III. Errors. Part IV: Index to introduction and Part I. Second Edition. Addicon-Wesley Publishing Co., Inc., Reading, Mass. 1962.
FRAHN,W.E., and R.H.LEMMER: Velocity dependent nuclear interaction. Nuovo Cim. X,5, 1564-1572 (1957).
FRAHN,W.E., and R.H.LEMMER: Non-static effects on individual nucleons in a spheroidal potential. Nuovo Cim.X, 6, 664-673 (1957).
FRIEDEN,B.R.: Evaluation, design and extrapolation methods for optical signals, based on use of the prolate functions. Progress in Optics 9, 311-407 (1971).
FRISCHBIER,R., and F.M.HESSELBARTH: Neues zur Theorie der Mathieufunktionen. Z.Angew. Math.Mech. 51, 485-487 (1971).
FUCHS,W.H.J.: On the eigenvalues of an integral equation arising in the theory of band-limited signals. J.Math.Analysis and Appl. 9, 317-330 (1964).
GAUDIN,M.: Sur la loi limite de l'espacement des valeurs propres d'une matrice aléatoire. Nucl.Phys. 25, 447-458 (1961).

GERMEY,K.: Die Beugung einer ebenen elektromagnetischen Welle an zwei parallelen unendlich langen idealleitenden Zylindern von elliptischem Querschnitt. Ann. Physik (7) 13, 237-251 (1964).
GOODRICH,R.F., and N.D.KAZARINOFF: Diffraction by thin elliptic cylinders. The Michigan Math.J. 10, 105-127 (1963).
GRAY,H.J., R.MERVIN and J.G.BRAINERD: Solutions of the Mathieu equation. Trans.Am. Inst.Elect.Engrs. 67, 429-441 (1948).
GUPTA,R.K.: A finite transform involving Mathieu functions and its application. Proc.Nat.Inst.Sci.India A 30, 779-795 (1964).
GUPTA,R.K.: A finite transform involving generalized prolate spheroidal wave functions and its applications. Indian J. Pure Appl.Math. 6, 432-443 (1975).
GUPTA,R.K.: A finite transform involving generalized prolate spheroidal wave functions and its applications. Proc.Indian Acad.Sci.Sect. A 85, 41-54 (1977).
GUPTA,R.K.: Generalized prolate spheroidal wave functions. Proc.Indian Acad.Sci. A 85, 104-114 (1977).
HALVORSEN,S.: On the characteristic exponents of the Mathieu equation. Norske Vid. Selsk.Forh. (Trondheim) 37, 13-18 (1964).
HANISH,S., R.V.BAIER, A.L.VAN BUREN and B.J.KING: Tables of spheroidal wave functions. Vols. 1-3, Prolate, m = 0,1,2; Vols. 4-6, Oblate, m = 0,1,2. Naval Res.Lab. Reports 7088-7093, 1970; Washington, D.C.
HANSEN,R.C.: Electromagnetic field solutions for rotational coordinate systems. Canad.J.Phys., 34, 893-895 (1956).
HATCHER,E.C., and A.LEITNER: Radiation from a point dipole located at the tip of a prolate spheroid. J.Appl.Phys. 25, 1250-1253 (1954).
HENKE,H.: Eine Methode zur Lösung spezieller Randwertaufgaben der Mathieuschen Differentialgleichung. Z.Angew.Math.Mech. 52, 250-251 (1972).
HERSCHBACH,D.R.: Tables of Mathieu Integrals for the Internal Rotation Problem. J.Chem.Phys., 27, 975 (1957).
HEURTLEY,J.C.: Hyperspheroidal functions - optical resonators with circular mirrors. pp. 367-375 in: Proc.Symp. on Quasi-Optics; ed.J.Fox.Polytechn.Press, Brooklyn, N.Y., 1964.
HEURTLEY,J.C.: A theoretical study of optical resonator modes and a new class of special functions, the hyperspheroidal functions. Ph.D.dissertation 1965, University of Rochester, Rochester, N.Y..
HOCHSTADT,H.: Special functions of mathematical physics. Athena Series: Selected Topics in Mathematics. Holt, Rinehart and Winston, New York, 1961.
HOCHSTADT,H.: Instability intervals of Hill's equation. Comm.Pure Appl.Math. 17, 251-255 (1964).
HÖHN,E.: Numerische Untersuchungen zu den von J.Dörr angegebenen Integralgleichungen. Z.angew.Math.Mech. 38, 175-179 (1958).
HONG,K.M., and J.NOOLANDI: Solution of the Smoluchowski equation with a Coulomb potential. I. General results. J.Chem.Phys. 68, 5163-5171 (1978).
JAWSON,M.A.: Limiting properties of Mathieu functions. Proc.Cambridge Philos.Soc. 53, 111-114 (1957).
JORNA,S.: Derivation of Green-type, transitional, and uniform expansions from differential equations. IV. Periodic Mathieu functions ce(z,h) and se(z,h) for large h. Proc.Roy.Soc.Lond.A 286, 366-375 (1965).
JORNA,S., and C.SPRINGER: Derivation of Green-type, transitional and uniform asymptotic expansions from differential equations. V. Angular oblate spheroidal wavefunctions $ps_n^r(\eta,h)$ and $qs_n^r(\eta,h)$ for large h. Proc.Roy.Soc.Lond. A 321, 545-555 (1971).
KALNINS,E.G., and W.MILLER,JR.: Lie theory and the wave equation in space-time. I. The Lorenz group. J.Mathem.Phys. 18, 1-16,(1977).
KALNINS,E.G., and W.MILLER,JR.: Lie theory and the wave equation in space time. 4. The Klein-Gordon equation and the Poincaré group. J.Mathem.Phys. 19, 1233-1246, (1978).
KARMAZINA,L.N.: On the asymptotics of spheroidal wave functions. Vyčisl.Mat. 5, 72-78 (1959).
KING,B.J., and A.L.VAN BUREN: A general addition theorem for spheroidal wave functions. SIAM J.Math.Anal. 4, 149-160 (1973).
KOMAROV,I.V., L.I.PONOMAREV and S.Ju.SLAVYANOV: Sferoidaln'ye i Kulonovskie Sferoidal'nye Funkcii (Spheroid- and Coulomb-Spheroidfunctions). Moskau 1976.

KOŠPARENOK,V.N., and V.P.ŠESTOPALOV: Partial inversion of the operator for summator functional equations with the kernel in the form of periodic Mathieu functions. Dokl.Akad.Nauk SSSR 239, 811-814 (1978) = Sov.Phys.Dokl. 23 (4), 235-237 (1978).
KRANK,W.: Über die Theorie und Technik des elliptischen Wellrohrhohlleiters. Dissertation Aachen 1964.
KRICKEBERG,K.: Über die asymptotische Darstellung der Aufspaltung von Paaren benchbarter Eigenwerte der Differentialgleichung der Sphäroidfunktionen. Z.angew. Math.Physik (ZAMP) 6, 235-238 (1955).
KURZ,M.: Fehlerabschätzungen zu asymptotischen Entwicklungen der Eigenwerte und Eigenlösungen der Mathieuschen Differentialgleichung. Dissertation Essen 1979.
LAMB,H.: On the oscillations of a viscous spheroid. Proc.Math.Soc. (London) 13, 51-66, 189-212 (1881).
LANDAU,H.J., and H.O.POLLAK: Prolate spheroidal wave functions, Fourier analysis and uncertainty - II.Bell Syst.Techn.J. 40, 65-84 (1961).
LANDAU,H.J., and H.O.POLLAK: Prolate spheroidal wave functions, Fourier analysis and uncertainty - III: The dimension of the space of essentially time - and band-limited signals. Bell System.techn.J. 41, 1295-1336 (1962).
LATTA,G.E.: Some differential equations of the Mathieu type, and related integral equations. J.Math. and Phys. 42, 139-146 (1963).
LAUCHLE,G.C.: Local radiation impedance of vibrating prolate spheroids. J.Acoust. Soc.Am. 51, 1106-1109 (1972).
LAUCHLE,G.C.: Radiation of sound from a small loudspeaker located in a circular baffle. J.Acoust.Soc.Am. 57, 543-549 (1975).
LEITNER,A., und J.MEIXNER: Simultane Separierbarkeit von verallgemeinerten Schwingungsgleichungen. Arch.Math. 10, 387-391 (1959).
LEITNER,A., und J.MEIXNER: Eine Verallgemeinerung der Sphäroidfunktionen. Arch.Math. 11, 29-39 (1960).
LEVY,B.R., and J.B.KELLER: Diffraction by a spheroid. Can.J.Phys. 38, 128-144 (1960).
LEVY,D.M., and J.B.KELLER: Instability intervals of Hill's equation. Comm.Pure Appl. Math. 16, 469-476 (1963).
LOTSCH,H.K.V.: The Fabry-Perot Resonator. Optik 28, 65-75, 328-345, 555-574 (1968/69), 29, 130-145, 622-623 (1969).
LOTSCH,H.K.V.: The confocal resonator system. Optik 30, 1-14, 181-201, 217-233, 563-576 (1969).
LOWAN,A.N.: Spheroidal wave functions, p.751-769 in: Handbook of Mathematical Functions. Ed. by M.Abramowitz and I.A.Stegun. Dover Pub., Inc.,New York, 1965.
LYTLE,R.J., and F.V.SCHULTZ: Prolate spheroidal antennas in isotropic plasma media. IEEE Trans.AP - 17, 496-506 (1969).
MATSUMOTO,T.: Note on the integral representations of Mathieu functions. Mem.Coll Sci., Univ. Kyoto, Ser. A 27, 133-137 (1952).
MEIXNER,J., und F.W.SCHÄFKE: Mathieusche Funktionen und Sphäroidfunktionen mit Anwendungen auf physikalische und technische Probleme. Springer Verlag Berlin/Göttingen/Heidelberg (1954).
MEIXNER,J., und F.W.SCHÄFKE: Eigenwertkarten der Sphäroiddifferentialgleichung. Arch.Math. 5, 492-505 (1954).
MEIXNER,J., and C.P.WELLS: Improving the convergence in an expansion of spheroidal wave functions. Quart.Appl.Math. 17, 263-269 (1959).
MEIXNER,J.: Einige Eigenschaften der Sphäroidfunktionen. Arch.Math. 20, 274-278 (1969).
MEIXNER,J., and SCHIU SCHE: Some remarks on the treatment of the diffraction through a circular aperture. Philips Res.Repts. 30, 232*-239* (1975).
MEIXNER,J.: Orthogonal polynomials in the theory of Mathieu functions. To be published.
MENNICKEN,R.: Entwicklungen analytischer Funktionen nach Produkten Whittakerscher Funktionen. Dissertation Köln 1963.
MENNICKEN,R.: Neue numerische Verfahren zur Berechnung des charakteristischen Exponenten der verallgemeinerten Mathieuschen Differentialgleichung $(1+2\gamma \cos 2x)y''(x) + (\lambda-2h^2 \cos 2x)y(x) = 0$ Arch.Rational Mech.Anal. 26, 163-178 (1967).
MENNICKEN,R., und A.SATTLER: Biorthogonalentwicklungen analytischer Funktionen nach Produkten spezieller Funktionen. Math.Z. 89, 1-29 (1965); 89, 365-394 (1965).
MENNICKEN,R., und A.SATTLER: Biorthogonalentwicklungen analytischer Funktionen nach Eigenlösungen linearer Differentialgleichungen. Math.Z. 93, 1-36 (1966).

MILES,J.W.: Asymptotic approximations for prolate spheroidal wave functions. Studies Appl.Math. 54, 315-349 (1975).
MILLER,JR.,W.: Symmetry, separation of variables, and special functions. pp.305-352 in:Theory and application of special functions. Ed.R.A. Askey.Acad.Press, New-York-San Francisco-London 1975.
MILLER,W.,JR.: Symmetry and separation of variables. Encyclopedia of Mathematics and its applications. Addison-Wesley Publishing Co., Reading, Mass.. - London-Amsterdam, 1977.
MÜLLER,H.J.W.: Asymptotic expansions of oblate spheroidal wave functions and their characteristic numbers. J.reine angew.Math. 211, 33-47 (1962).
MÜLLER,H.J.W.: On asymptotic expansions of Mathieu functions. J.reine angew.Math. 211, 179-190 (1962).
MÜLLER,H.J.W.: Asymptotic expansions of prolate spheroidal wave functions and their characteristic numbers. J.reine angew.Math. 212, 26-48 (1963).
MÜLLER,H.J.W.: Asymptotische Entwicklungen von Sphäroidfunktionen und ihre Verwandtschaft mit Kugelfunktionen. Z.angew.Math.Mech. 44, 371-374 (1964).
MÜLLER,H.J.W.: Über asymptotische Entwicklungen von Sphäroidfunktionen. Z.angew.Math. Mech. 45, 29-36 (1965).
NEUBERGER,H.: Semiclassical calculation of the energy dispersion relation in the valence band of the quantum pendulum. Phys.Rev.D, 17, 498-506 (1978).
PAL'CEV,A.A.: The computation of spheroidalfunctions and their first derivatives on a computer. I. Vescī Akad.Navuk BSSR Ser. Fiz.-Mat. Navuk 1969 no. 1, 19-25.
PALERO,R.S.B.: On the differential equation

$$\frac{d^2y}{dx^2} + \left[\frac{a_0 + a_1 \cos 2x}{b_0 + b_1 \cos 2x} - \frac{m(m-1)}{\sin^2 x} - \frac{n(n-1)}{\cos^2 x}\right] y = 0$$

Mem.Mat.Inst. "Jorge Juan" nr. 18, (1956), i + 145 pp.
PARODI,M.: Equations de Mathieu et équations intégrales de Volterra. C.R.Acad.Sci. Paris 243, 1006-1007 (1956).
PARODI,M.: Equations intégrales et équations du type de Mathieu. J.Math.Pures Appl. (9) 37, 45-54 (1958).
PIEFKE,G.: Asymptotische Näherungen der modifizierten Mathieuschen Funktionen. Z.angew.Math.Mech. 44, 315-327 (1964).
PIPES.L.A.: Matrix solution of equations of the Mathieu-Hill type. J.Appl.Phys. 24, 902-910 (1953).
PROCENKO,V.S., and V.G.PROCENKO: A certain case of an addition theorem for Mathieu functions. Dopovīdī Akad.Nauk Ukrain.RSR Ser.A 1972, 808-810, 861.
RAND,R.H.: Torsional vibrations of elastic prolate spheroids. J.Acoust.Soc.Am. 44, 749-751 (1968).
RAUH,H., and D.SIMON: On the diffusion process of point defects in the stress field of edge dislocations. phys.stat.sol.(a) 46, 499-510 (1978).
REMBOLD,B.: Asymptotische Näherungen der modifizierten Mathieuschen Funktionen für komplexe Eigenwertparameter $h = |h|(1+j)/\sqrt{2}$. Z.angew.Math.Mech. 53, 783-789 (1973).
RHODES,D.R.: On some double orthogonality properties of the spheroidal and Mathieu functions. J.Math. and Phys. 44, 52-65 (1965).
RHODES, D.R.: On the spheroidal functions. J.Res.Nat.Bur.Standards Sect. B 74, 187-209 (1970).
RHODES,D.R.: On a third kind of characteristic numbers of the spheroidal functions. Proc.Nat.Acad.Sci. USA 67, 351-355 (1970).
RIBIERE-MICHAUD,F.: Le calcul numérique des fonctions de Mathieu en tant que solutions d'équations intégrales. Ž.Vyčisl.Mat.i.Mat.Fiz. 9, 704-707 (1969).
ROBIN,L.: Fonctions sphériques de Legendre et fonctions sphéroidales. I, 1957; II, 1958; III, 1959. Gauthier-Villars, Paris.
ROBIN,L.: Diffraction d'une onde électromagnetique plane par un cylindre elliptique d'axe parallèle à l'onde. Extension à une bande plane et à une fente à bords parallèle dans un plan. Ann.Inst. Henri Poincaré, A, 3, 183-194 (1965).
ROBIN,L.: Sur un cas particulier de l'équation sphéroïdale. C.R.Acad.Sci., Paris 262, 496-498 (1966).
ROBIN,L.: Sur la définition et sur l'expression de la fonction sphéroidale de seconde espèce $Qs_n^m(z;\gamma)$. C.R.Acad.Sci. Paris Sér. A-B 262, A 566 - A 568 (1966).

ROBIN,L.: Quelques résultats sur les fonctions sphéroidals avec tables numériques de celles-ci. Revue du Centre d'Etudes Théoriques de la Détection et des Communications (CETHEDEC) 6ème année, 1969, Numéro Spécial.

ROBINSON,P.D.: Wave Functions for the Hydrogen Atom in Spheroidal Coordinates. II. Interaction with a Point Charge and with a Dipole. Proc.Phys.Soc., 71, 828- (1958).

RUBIN,H.: Anecdote on power series expansions of Mathieu functions. J.Math. and Phys. 43, 339-341 (1964).

SAERMARK,K.: A note on addition theorems for Mathieu functions. J.Angew.Math.Phys. 10, 426-428 (1959).

SATTLER,A.: Entwicklungen analytischer Funktionen in Kapteynsche Reihen zweiter Art. Dissertation Köln 1963.

SCHÄFKE,F.W.: Ein Verfahren zur Berechnung des charakteristischen Exponenten der Mathieuschen Differentialgleichung. Numer.Math. 3, 30-38 (1961).

SCHÄFKE,F.W.: Reihenentwicklungen analytischer Funktionen nach Biorthogonalsystemen spezieller Funktionen. Math.Z. 74, 436-470 (1960); 75, 154-191 (1961); 80, 400-442 (1963).

SCHÄFKE,F.W., und H.GROH: Zur Berechnung der Eigenwerte der Mathieuschen Differentialgleichung. Numer.Math. 4, 64-67 (1962).

SCHÄFKE,F.W., und H.GROH: Zur Berechnung der Eigenwerte der Sphäroiddifferentialgleichung. Numer.Math. 4, 310-312 (1962).

SCHÄFKE,F.W., und A.SCHNEIDER: S-hermitesche Rand-Eigenwert-Probleme. Math.Z. 162, 9-26 (1965); 165, 236-260 (1966); 177, 67-96 (1968).

SCHMIDT,D., and G.WOLF: A method of generating integral relations by the simultaneous separability of generalized Schrödinger equations. SIAM J.Math.Analysis 10, 823-838 (1979).

SCHNEIDER,A.: Eine Verallgemeinerung der Kapteynschen Reihen. Math.Z. 83, 212-236 (1963).

SHARMA,S.D.: Expansion of Kampé de Fériet function in terms of generalized prolate spheroidal wave function and its applications. Indian J. Pure Appl.Math. 9, 764-770 (1978).

SHARMA,S.D., and R.K.GUPTA: On double integrals involving generalized H-function and spheroidal functions. Acta Ci.Indica 3, 346-352 (1977).

SHARPLES,A.: Uniform asymptotic forms of modified Mathieu functions. Quart.J.Math. and Appl.Math. 20, 365-380 (1967).

SHARPLES,A.: Uniform asymptotic expansions of modified Mathieu functions. J.reine angew.Math. 247, 1-17 (1970).

SIDMAN,R.D.: Scattering of acoustical waves by a prolate spheroidal obstacle. J.Acoust.Soc.Am. 52, 879-883 (1972).

SILBIGER,A.: Radiation from circular pistons of elliptical profile. J.Acoust.Soc. Am. 33, 1515-1522 (1961).

SILBIGER,A.: Comments on "Sound radiation from prolate spheroids". J.Acoust.Soc.Am. 33, 1630 (1961).

SILBIGER,A.: Scattering of sound by an elastic prolate spheroid. J.Acoust.Soc.Amer. 35, 564-570 (1963).

SINHA,B.B., and R.H.MACPHIE: On the computation of the prolate spheroidal radial functions of the second kind. J.Math.Phys. 16, 2378-2381 (1975).

SIPS,R.: Représentation asymptotique des fonctions de Mathieu et des fonctions sphéroidales. II. Trans.Amer.Math.Soc. 90, 340-368 (1959).

SIPS,R.: Nouvelle méthode pour le calcul de l'exposant caractéristique de l'équation de Mathieu-Hill.Bull.Acad.roy.Belgique, Cl.Sc. 51, 191-206 (1965).

SIPS,R.: Représentation asymptotique de la solution générale de l'équation de Mathieu-Hill. Bull.Acad.roy. Belgique.Cl.Sc. 51, 1415-1446 (1965).

SIPS,R.: Répartition du courant alternatif dans un conducteur cylindrique de section elliptique. Bull.Acad.roy.Belgique,Cl.Sc. 53, 861-878 (1967).

SIPS,R.: Répartition du courant alternatif dans un conducteur cylindrique de section presque circulaire. Bull.Acad.roy.Belgique,Cl.Sc. 53, 1279-1290 (1967).

SIPS,R.: Quelques intégrales définies discontinues contenant des fonctions de Mathieu. Bull.Acad.roy.Belgique,Cl.Sc. 56, 475-491 (1970).

SLAVJANOV,S.Ju.: Asymptotic representations for prolate spheroidal functions. Ž.Vyčisl.Mat.i.Mat.Fiz. 7, 1001-1010 (1967).

SLEPIAN,D.: Some asymptotic expansions for prolate spheroidal wave functions. J.Math. and Phys. 44, 99-140 (1965).

SLEPIAN,D.: Prolate spheroidal wave functions, Fourier analysis and uncertainty-IV: Extensions to many dimensions; generalized prolate spheroidal functions. Bell Syst. Tchn.J. 43, 3009-3057 (1964).
SLEPIAN,D., and H.O.POLLAK: Prolate spheroidal wave functions, Fourier analysis and uncertainty - I. Bell Syst.Techn.J. 40, 43-64 (1961).
SLEPIAN,D., and E.SONNENBLICK: Eigenvalues associated with prolate spheroidal wave functions of zero order. Bell Syst.Techn.J. 44, 1745-1759 (1965).
SMITH,J.J. A method of solving Mathieu's equation. Trans.Amer.Inst.Elect.Engrs I, 74, 520-525 (1955).
STRATTON,J.A., P.M.MORSE, L.J.CHU, J.D.C.LITTLE and F.J.CORBATO: Spheroidal wave functions. Wiley, New York, 1956.
TABLES RELATING to Mathieu Functions. Characteristic values, coefficients, and joining factors. Second Edition. National Bureau of Standards. Applied Mathematics Series, Vol. 59, 1969, U.S.Govt.Printing Office, Washington,D.C. 20402.
TOMPKINS,Jr., D.R.: Viscous-fluid reaction to a torsionally oscillating spheroid: Linearized steady-state spheroidal function solution. Nuovo Cim. 21 B, 36-46 (1974).
TROESCH,B.A.: Elliptical membranes with smallest second eigenvalue. Math.Comp. 27, 767-772 (1973).
TROESCH,B.A., and H.R.TROESCH: Eigenfrequencies of an elliptic membrane. Math.Comp. 27, 755-765 (1973).
VAN BUREN,A.L.: Acoustic radiation impedance of caps and rings on prolate spheroids. J.Acoust.Soc.Am. 50, 1343-1356 (1971).
VAN BUREN,A.L.: Comments on'local radiation impedance of vibrating prolate spheroids'. J.Acoust.Soc.Am. 53, 1744-1746 (1973).
VAN BUREN,A.L., R.V.BAIER, S.HANISH and B.J.KING: Calculation of spheroidal wave functions. J.Acoust.Soc.Am. 51, 414-416 (1972).
VAN BUREN,A.L., and B.J.KING: Acoustic Radiation from two spheroids. J.Acoust.Soc. Amer. 52, 364-372 (1972).
VAN BUREN,A.L., B.J.KING, R.V.BAIER, S.HANISH: Tables of angular spheroidal wave functions. Vol. 1, Prolate, m = 0; Vol. 2, Oblate, m = 0. Naval Res.Lab.Reports, June 30, 1975; Washington,D.C.
VAN DER HEIDE,H.: Stabilisierung durch Schwingungen. Philips techn.Rundschau 33, 361-373 (1973/74).
VELO,G., and J.WESS: A solvable quantum-mechanical model with nonlinear transformation laws. Nuovo Cimento 1 A, 177-187 (1971).
VILLE,J.A., and J.BOUZITAT: Note sur un signal de durée finie et d'énergie filtrée maximum. Cables & Trans. 11, 102-127 (1957).
VOLKMER,H.: Integralrelationen mit variablen Grenzen für spezielle Funktionen der mathematischen Physik. Dissertation Konstanz, 1979.
WAGENFÜHRER,E.: Ein Verfahren höherer Konvergenzordnung zur Berechnung des charakteristischen Exponenten der Mathieuschen Differentialgleichung. Numer.Math. 27, 53-65 (1976).
WALL,D.J.N.: Circularly symmetric Green tensors for the harmonic vector wave equation in spheroidal coordinate systems. J.Phys.A 11, 749-757 (1978).
WIDOM,B.: Inelastic molecular collisions with a Maxwellian interaction energy. J.Chem.Phys. 27, 940-952 (1957).
WIDOM,H.: Asymptotic behavior of the eigenvalues of certain integral equations II, Arch.Rational Mech.Anal. 17, 215-229 (1964).
WINTNER,A.: A note on Mathieu's functions. Quart.J.Math.Oxford Ser. (2) 8, 143-145 (1957).
WOLF,G.: Additionstheoreme Mathieuscher Funktionen. Dissertation Cologne 1969.
WOLF,G.: Integralrelationen Mathieuscher Funktionen. Arch.Rational Mech.Anal. 45, 134-142 (1972).
YAO,K., and J.B.THOMAS: On bandlimited properties of Fourier transform pairs of some special functions. Proc.Third Annual Allerton Conf. on Circuit and System Theory. Univ. Illinois, Urban, Ill., 1965.
YEH,C.: The diffraction of waves by a penetrable ribbon. J.Math.Phys. 4, 65-71 (1963).
YEH,C.: Scattering of acoustic waves by a penetrable prolate spheroid, I. Liquid prolate spheroid. J.Acoust.Soc.Am. 42, 518-521 (1967).
YEH,C.: Scattering by liquid-coated prolate spheroids. J.Acoust.Soc.Am. 46, 797-801, (1969).